ISBN 978-0-265-26154-5
PIBN 10897582

historic structure report

HISTORIC STRUCTURE REPORT

THREE SISTERS LIGHTHOUSES
CAPE COD NATIONAL SEASHORE
MASSACHUSETTS

by
A. Berle Clemensen
and
William W. Howell
with
H. Thomas McGrath
and
Elayne Anderson

January 1986

U.S. Department of the Interior / National Park Service

ACKNOWLEDGEMENTS

My thanks are extended to the Cape Cod National Seashore personnel who took the time to aid my research. I am also grateful to those individuals in the National Archives and the United States Coast Guard Library who provided materials.

A. Berle Clemensen
Denver, Colorado
November 14, 1984

I wish to thank the Superintendent and staff of Cape Cod National Seashore, especially the Maintenance and Interpretation divisions for their assistance with the physical work of investigating the structures and with research in the park files respectively.

I also want to recognize the significant contributions of H. Thomas McGrath and Elayne Anderson to the field research, measured drawings, and proposed work aspects of this report and for their helpful advice and revision comments on the text.

William W. Howell
Denver, Colorado
November 14, 1984

TABLE OF CONTENTS

LIST OF ILLUSTRATIONS

Figure

LIST OF MAPS

Map

LIST OF TABLES

Table

ADMINISTRATIVE DATA SECTION

Management Information

The historic Nauset Beach lighthouses, commonly known as the Three Sisters, are tapering circular towers of wood frame construction with white shingle siding. They are also known as the Beacon (with its lantern intact) and the Twins from their periods of use as parts of summer cottages. This report describes the Three Sisters as they appeared when inspected in February and June 1983. After June 1983, the wings of the Beacon (added for summer cottage purposes) were demolished and the Beacon was moved on September 13, 1983 to a temporary storage location near the Twins.

Proposed Use

The National Park Service intends to move the Three Sisters to a new site, restore their linear arrangement, and restore the lighthouses for interpretation to park visitors.

Planning Background

1. The Master Plan for Cape Cod National Seashore was approved in October 1970.

2. Topographical surveys of the Nauset Beach and Doane Rock areas were conducted by Sewall Co. and are filed as drawings numbered 609/41,034 and 609/41,036 at the Denver Service Center.

3. The Historic Resource Study for Cape Cod National Seashore was approved in February 1979.

4. An Assessment of Alternatives was prepared in July 1978 and a supplement to it was prepared in March 1980.

5. The Finding of No Significant Impact is dated August 1980.

6. The Development Concept Plan for the Eastham Area was approved in September 1981. The treatment of the Three Sisters falls under the general guidance of this document.

7. Comprehensive Design for the development of the Nauset Light Beach Complex was approved in September 1982.

8. The addendum to the task directive that has governed the production of this report was approved in May 1983.

The history and architecture components of this historic structure report were prepared concurrently. The project historian, A. Berle Clemensen, has done a thorough job of research at the usual (and some out of the way) repositories of documentary records.

Information relating directly to the structures at the Nauset Beach Light Station has been extracted from the history component and presented as a "Brief History of Physical Changes" and a "Chronology of Structures at Nauset Beach Light" in the architecture component.

In the course of our architectural research, we have come across two possible avenues for further research. The stencil "B. James & Co." found on one of the lantern deck beams in the North Twin suggests research into the history of commercial building supply companies on Cape Cod to learn more about the suppliers of materials for the lighthouses. In the course of our effort to locate original drawings for the lighthouses or their lanterns, we learned that the lanterns were not built in accordance with any of the standard designs that had been used by the Lighthouse Service since 1873. A thorough study of America's Lighthouses: Their Illustrated History Since 1716 by Francis Ross Holland, Jr. identified a similar lantern atop the Aransas Pass Lighthouse in Texas. Mr. Holland's 1976 publication, The Aransas Pass Light Station: A History, revealed that Hayward, Bartlett and Company, a Baltimore iron foundry, had fabricated the lantern for the Aransas Pass Lighthouse. Telephone calls to the present owner of the light station, Mr. Charles C. Butt, and to the keeper of the light confirmed details of construction of the two lanterns to be identical. The details that were checked were: the three piece cast iron sill, the shape of the glazing bars, and the number and size of plate glass lights. Further research into the history of Hayward, Bartlett and Company might prove conclusively that the lantern now on the Beacon was indeed fabricated by them. It would certainly be of academic value and it might show that the Three Sisters are more significant historically than is now thought.

Proposed Treatment and Justification

Because the Three Sisters are included in a thematic nomination of Massachusetts lighthouses to the National Register of Historic Places, they are subject to the requirements of the Advisory Council on Historic Preservation's "Procedures for the Protection of the Historic and Cultural Environment" (36 CFR Part 800), and the Regional Director shall, in consultation with the State Historic Preservation Officer, apply the Advisory Council's "Criteria of Effect" (Section 800.8) and "Criteria of Adverse Effect" (Section 800.9) and afford the Advisory Council the opportunity to review and/or comment on any treatment proposal.

The approved development concept plan (DCP) for the Eastham area proposed the following treatment for the Three Sisters

> The Three Sisters of Nauset (the Beacon and Towers) will be resited on a location on the west side of Ocean View Drive approximately midway between Doane and Cable roads on what is presently town land. Approximately 2 acres will be acquired through exchange, purchase, or donation to provide a setting for the lighthouses and a 15 car, short term (30 minutes) parking area.

With the Beacon on the south, all three lighthouses will be set approximately 150 feet appart from each other. All will receive exterior restoration in accordance with an historic structure report. Interiors will be stabilized except for the Beacons, which may be restored if sufficient information becomes available.

An interpretive wayside will provide the history and significance of these structures. As staffing permits, the interior of the Beacon will be open for public viewing and interpretation.[1]

With minor changes for historical accuracy, this proposed treatment is presented as the restoration alternative in this report. Any deviation from the actions proposed in the DCP will require compliance as stated in this section. However, a PMOA is in effect which covers the restoration alternative and would obviate the need to return to the SHPO and ACHP under the 36CFR800 regulations cited.

Recommended Treatment for Materials Collected in Preparing Report

Paint samples and a piece of shingle from the Beacon will be forwarded to the North Atlantic Historic Preservation Center for inclusion in their collection. A short length of cornice molding from the Beacon doorway will be returned to Cape Cod National Seashore to be held by them for reinstallation during the restoration project.

---

1. Development Concept Plan and Transportation Analysis, Eastham Area, Cape Cod National Seashore.

PHYSICAL HISTORY AND ANALYSIS SECTION

HISTORY COMPONENT

Significance

The significance of the Three Sisters is related to their role in coastal navigation, in the history of lighthouses, in local history, and as landmarks (in the literal sense as objects by which travelers, whether on land or sea, may orient themselves). Although the Twins lost their lights and lanterns some time between 1910 and 1918 and the Beacon has been out of service since 1923, together they may constitute the only surviving set of triple lights in the United States. The process of phasing out and dismantling multiple lights in separate structures at one location (such as the three lights at Holme's Hole in New England) began in the late 1850's. Furthermore, to the best of our knowledge, the lantern on the Beacon is one of two of its type in existence and the only one in federal ownership.

General Background of the Lighthouse Establishment

By an act of August 7, 1789, the United States government accepted cession of the title to coastal lighthouses from the several states and agreed to maintain them. The states ceded their lights between 1789 and 1797, with Massachusetts turning its eight lights over to the national government in 1790. Lighthouse administration was placed within the Treasury Department where the secretary had direct oversight until May 8, 1792, when control was vested with the newly-created commissioner of revenue. On April 6, 1802, the office of the commissioner of revenue was abolished and the secretary of the treasury again resumed direct management. Eleven years later the office of the commissioner of revenue was re-established with supervision of the lighthouses as one of the duties. An act of December 23, 1817, however, eliminated the office of commissioner of revenue for a second time as of January 1, 1820. On the latter date, control of the Lighthouse

Establishment came under the province of the fifth auditor of the treasury.[1]

Stephen Pleasonton, the fifth auditor, controlled the lights for 32 years until the Lighthouse Board was created on October 9, 1852. Pleasonton, an accountant by training, had little knowledge of lights or their operation. His main concern was keeping costs below governmental appropriations so that he could return money to the treasury each fiscal year. Early in his administration, Pleasonton came to rely upon Winslow Lewis, a former Massachusetts sea captain, who advised him on the needs of the Lighthouse Establishment.[2]

In 1810, Winslow Lewis developed a version of the Argand lamp and parabolic reflector for use in lighthouses. When his near parabolic device proved successful after testing in the Boston light, Lewis offered to sell the patent to the United States government and refit each of the existing 49 lighthouses with his contrivance. Congress accepted the offer in 1812. Thus, Lewis' lamp became the accepted illuminant until conversion to the Fresnel lens system in the 1850s. It was through this affiliation that Stephen Pleasonton came to know Winslow Lewis.[3]

Pleasonton executed the orders of Congress to construct lights and fixed their location, if necessary. Actually, the collector of customs nearest the light was designated local superintendent. He was the individual who approved the exact site for a lighthouse and hired a construction overseer to ensure that the contractor built the structure according to specifications. During most of Pleasonton's tenure, the Lighthouse Establishment did not employ architects or engineers to design

---

1. Arnold B. Johnson, The Modern Light-House Service (Washington, D.C.: Government Printing Office, 1890), pp. 13-4; George Weiss, The Lighthouse Service: Its History, Activities and Organization, (Baltimore: The Johns Hopkins Press, 1926), pp. 2-4.

2. Johnson, The Modern Light-House Service, pp. 14-6; Francis Ross Holland, Jr., America's Lighthouses: Their Illustrated History Since 1716 (Brattleboro, Vermont: The Stephen Green Press, 1972), pp. 16, 18.

lights. Instead, according to Lewis' nephew, Pleasonton had Lewis provide the description and specifications for lights. In addition, Lewis often received the contract to build a lighthouse by underbidding others. As a result, to make money, he often hurriedly constructed the lights without regard to the specification that he had authored.[4]

In the first 60 years of the national government's control of lighthouses, the location and construction of lights usually evolved from local requests made to Congress. These requests grew in number with each passing year. The money approved for lighthouse construction became so large that, in the March 3, 1837, appropriation, Congress called for a board of naval commissioners to examine the proposed sites and determine the necessity for a light at each one.[5]

The First Nauset Beach Lights

Among those new lighthouses for which Congress made an appropriation in the March 3, 1837 act, $10,000 was awarded for three designated lights at Nauset Beach on Cape Cod. The request for these lights evolved from a study made by the Boston Marine Society. In February 1833 this organization voted to establish a three-member committee that would examine and report on the need for an additional light on Cape Cod. The committee evidently did not actively pursue its investigation, for in early 1836 the society received a letter from 21 Eastham area residents who asked for support to establish a light at

---

3. Holland, America's Lighthouses, pp. 14-16. Argand, a Frenchman, invented the Argand lamp and parabolic reflector in 1781.

4. Johnson, The Modern Light-House Service, p. 14; "Report of I.W.P. Lewis, Civil Engineer, Made By Order of Hon. W. Forward, Secretary of the Treasury on the Condition of the Light-Houses, Beacons, Buoys, and Navigation, upon the Coasts of Maine, New Hampshire, and Massachusetts, in 1842," HR Doc. No. 183, February 24, 1843, 27th Con., 3rd Sess., p. 18, hereafter cited as "Report of I.W.P. Lewis."

5. Johnson, The Modern Light-House Service, p. 15.

Nauset Beach. Upon receipt of the letter, the committee examined the site and reported favorably in March 1836. As a result, the Boston Marine Society petitioned Congress to appropriate $10,000 to light that beach. It recommended that three lighthouses be placed there to prevent confusion with other lights (the existing Highland light and the two lighthouses at Chatham). In addition, the society suggested that the beacons be about 15 feet high.[6]

The rationale for lighting Nauset Beach was not one of providing top-of-the-line lighthouses. Ocean-going vessels, that did not come close to shore at Nauset Beach, could observe the Highland and Chatham lights and thus chart a course beyond the Nauset shoals. Only those coastal and fishing vessels that hugged the shore encountered problems. The curving cape shoreline prevented craft from viewing either the Highland or Chatham lights when in the vicinity of Nauset Beach. Thus, without a guiding light, coastal vessels were subject to grounding on the Nauset bars. As a result, the need for a light(s) at Nauset served only an intermediate purpose--signaling ships of the local danger until they could view the Highland or Chatham lights.[7]

As stated previously, Congress approved the funds for the Nauset lights contingent upon an inspection of the site by a naval officer. The appropriation act read, "for three small light-houses on Nanset [sic] beach, Cape Cod, fifteen feet high, ten thousand dollars." Captain John Percival of the navy inspected the site and approved it as a necessary location for lights. He drove stakes into the ground on a line parallel with the coast at the points he selected for each lighthouse. David

---

6. William A. Baker, A History of the Boston Marine Society, 1742-1967, (Boston: Boston Marine Society, 1968) p. 134. The Boston Marine Society's membership was restricted to individuals who then commanded ships or those who had in the past.

7. Winslow Lewis to Stephen Pleasonton, Fifth Auditor of the Treasury Department, April 6, 1842, Box 7, Light-House Superintendent, Boston 1826-44, Records of the United States Coast Guard, Record Group 26, National Archives, Washington, D. C.

Henshaw, the Boston customs collector who acted as the area superintendent of lights, purchased approximately 5 acres of land at the Nauset site from Benjamin H.A. Collins of Eastham on September 14, 1837.[8]

In the early 19th century, the lighthouses of Massachusetts were constructed of either rubble stone, brick, or wood. Walls were usually 3 feet thick at the base, tapering to 2 feet at the top. The structures ranged up to 50 feet in height. A dome was placed inside the top which contained a square opening near the spring line to give entrance to the lantern. A flat, slab-stone roof, 4 inches thick, was placed over the dome and projected over the tower walls from 6 to 12 inches. Lanterns were attached to the towers by imbedding the lower ends of their iron angle posts into the walls some 3 to 4 feet.[9]

Winslow Lewis, as low bidder, signed the contract on May 26, 1838, to build the three Nauset lights, a keeper's dwelling, an outhouse, and well for $6,549. The contract specifications, which had probably been written by Lewis, called for three, round brick towers each 15 feet high from the ground to the top of the walls. They appeared to be patterned after the common type found in Massachusetts. Each was to have a base diameter of 16 feet, tapering to a top diameter of 9 feet. The thickness of the base walls was to be 3 feet, while graduating to 20 inches at the top. An arch was placed at the interior top with a 4-inch thick soapstone roof or deck projecting 18 inches beyond the wall. Granite stairs were to lead from the ground to a platform 6 feet from the top. At

---

8. "An Act Making Appropriations for Building Light-Houses, Light-boats, Beacon-lights, Buoys, and Dolphins for the Year 1837," _United States Statutes At Large_, 5 (Boston: Charles C. Little and James Brown, 1850), p. 182; "Affidavit of David Bryant, Carpenter, and Superintendent of Construction at the Nauset Light-Houses, Erected in 1838," Report of I.W.P. Lewis, p. 29; David Henshaw to Stephen Pleasonton, September 19, 1837, Box 5, Light-House Superintendent, Boston, 1834-48, Records of the United States Coast Guard, Record Group 26, National Archives, Washington, D.C.

9. Report of I.W.P. Lewis, pp. 25-26.

that point, an iron ladder led to the lantern's entrance. Each tower was to have two windows composed of twelve, 8- by -10-inch lights, and one, 6- by 3-foot door. Iron octagon lanterns were to top the towers. Their 1-1/2-inch square posts were to be imbedded 4 feet into the walls. In addition, the contract called for a 34- by 20-foot one-story brick house with a 16- by 14-foot kitchen addition, a 4- by 5-foot wood outhouse, and a well.[10]

Lewis' construction crew of four masons, two carpenters, three laborers, and a cook arrived in June. They proceeded to relocate the middle and south lights back from the staked sites to lower ground. The reason for such action was to place the three lights at the same elevation and thus save building materials, otherwise the northernmost light would have had to be twice the height to make it level with the other two. During construction, which took only 38 consecutive days, the workmen often ignored the building specifications. Some points of the contract, however, were so loosely worded that it was left to the contractor to interpret them in an honest manner. David Bryant, the construction supervisor, felt that the work had been so poorly accomplished that he refused to sign the completion certificate presented by Winslow Lewis which stated that the terms of the contract had been honorably fulfilled. Instead, he referred Lewis to George Bancroft, the superintendent in Boston. He also notified Bancroft of his action. Bancroft, however, told Bryant to sign the completion certificate. When he hesitated to place his signature on a questionable document, Bancroft informed him that Stephen Pleasonton had accepted Lewis's word that the construction had been properly finished. As a result, Bryant signed the certificate.[11]

Stephen Pleasonton's ignorance of appropriate construction methods and the trust he had placed in Winslow Lewis over the years did not benefit the Light House Establishment. The two men's relationship,

---

10. Ibid., pp. 31-33.

11. Report of I.W.P. Lewis, pp. 29-30. George Bancroft later became a noted nineteenth century American historian.

however, was mutually advantageous. For his part, Lewis received many contracts to construct lighthouses and, although he bid low, he made money by cutting corners. At the same time, Pleasonton's parsimonious, accountant nature was satisfied, for as in the case of the Nauset lights, he could return $3,451 of the appropriated construction money to the treasury and thereby appear to be carefully handling the public money.

Pleasonton, however, was not without his share of critics about the manner in which he operated the Lighthouse Establishment. Sometimes these fault-finders would succeed in presenting their case to Congress. One of the first occasions occurred in 1838. In that year Pleasonton's opponets succeeded in getting a congressional bill which called for the creation of six Atlantic coast and two Great Lakes lighthouse districts with naval officers detailed to inspect the lights in each and to report on their condition. Massachusetts became part of the second district and Lieutenant Edward Carpenter was assigned as the inspector. He promptly traveled to each facility. When he arrived at Nauset Beach several months after the completed construction, he found that the lights had yet to be lit for lack of a fuel delivery. Carpenter recognized that Nauset Beach needed to be illuminated because the sand bars had caused the destruction of many vessels, but he could not understand why three lights had been built when one would suffice. He concluded, "I cannot believe that the Government will consent to consume 900 gallons of oil, when 300 or 360 will answer every purpose. Accordingly, I shall recommend the conversion of these lights into a single revolving red light." While at the site, Carpenter noted that Winslow Lewis had installed ten lamps in each lantern.[12]

Although Pleasonton weathered the findings of the naval officers' inspection reports which indicated poor management, his critics continued

---

12. Johnson, The Modern Light-House Service, p. 16; report of Lieutenant Edward W. Carpenter, U.S.N., New York, November 1, 1838, Box 2, Clipping File, Edgartown, Massachusetts 2nd District - Barnegat, New Jersey, 3rd District, Records of the United States Coast Guard, Record Group 26, National Archives, Washington, D.C. The near parabolic reflectors in each of Lewis' ten lamps measured thirteen and a half inches.

to complain about the lighthouse system. As a result, in May 1842, the secretary of the treasury appointed Isaiah W.P. Lewis, a civil engineer of unquestioned repute, to inspect the lights of Maine, New Hampshire, and Massachusetts and to report on their condition and management. Lewis, the nephew of Winslow Lewis, produced a document that did not flatter the Lighthouse Establishment or his uncle. He focused on the Nauset Beach lights as a typical example of fraud practiced by contractors. The three Nauset Beach lighthouses, he wrote, were constructed on sand without foundations. The contractor used inferior lime in the mortar. Bricks were laid at random with no regard to forming a bond. There were 24 courses of stretchers for every course of headers. The lower windows in each of the towers had been boarded up because blowing gravel, during storms, kept breaking the glass. Lewis recommended that there be only one lighthouse at Nauset with a revolving light flashing every 1-1/2 minutes.[13]

Isaiah Lewis' report caused a stir with Stephen Pleasonton and his uncle. Pleasonton complained that the account had unfairly and grossly misrepresented him. Winslow Lewis denounced his nephew as an incompetent who was wholly incapable of judging proper lighthouse construction methods.[14]

The secretary of the treasury agreed with I.W.P. Lewis and recommended to Congress that an engineer be appointed who would devote full time to the regulation of details of the lighthouse system. Congress adjourned without taking any action on the matter.[15]

It was not just the poorly constructed lighthouses that received criticism but also the lighting system. The lamp, which Winslow Lewis

---

13. Johnson, The Modern Light-House Service, p. 18; report of I.W.P. Lewis, pp. 17, 28-29, 179.

14. Johnson, The Modern Light-House Service, p. 18; Holland, America's Lighthouses, p. 33.

invented and for which he had become the sole supplier to the Lighthouse Establishment, had been eclipsed in value with the introduction of the Fresnel lens in France in the 1820s. His influence with Stephen Pleasonton, however, kept his lamp in use as the principal source of lighting until the 1850s. The lamps proved to be flimsily constructed, for the reflectors could be easily bent, thereby reducing their effectiveness. In addition, the silver plating on the reflectors was easily removed with the abrasive cleaning material supplied to the keepers. When complaints about the lamps surfaced by 1838, Congress ordered Pleasonton to import two Fresnel lenses for experimental purposes, but the fifth auditor stubbornly defended the Lewis lamp. Among others, I.W.P. Lewis condemned his uncle's lamps in his 1843 inspection report. Finally, the critics became so numerous that in March 1851, Congress authorized Pleasonton to begin the importation of Fresnel lenses in a move to convert the illuminating equipment of America's lighthouses.[16]

The same 1851 act authorized and required the treasury secretary to assemble a board of inquiry to investigate every aspect of the Lighthouse Establishment. After an exhaustive study, the board, as had I.W.P. Lewis before it, found every phase of the Lighthouse Establishment to be faulty. The investigative board recommended that the administration of lighthouses be removed from the fifth auditor and that the responsibility be placed under a 9-member lighthouse board whose membership would be mostly army engineers and naval personnel. It proposed the creation of 12 districts, each with its own inspector who had an engineering background. The board asked that lighthouses be classified, after the French model, into six orders with orders one through three designated as coastal lights and four through six commissioned as harbor lights. Finally, they advised the replacement of the Lewis lamps and reflectors

---

15. Johnson, The Modern Light-House Service, p. 19.

16. Johnson, The Modern Light-House Service, p. 20; Holland, America's Lighthouses, p. 18; report of I.W.P. Lewis, pp. 51-54.

with the Fresnel lenticular system. Congress acted on the recommendations and approved the necessary legislation on August 31, 1852, to establish a Lighthouse Board.[17]

The Lighthouse Board, established under the secretary of the treasury, was organized on October 9, 1852. One of its first acts was to begin the process by which each lighthouse would receive the Fresnel lens system. Along with the nation's other lights, the ones at Nauset Beach came under the board's watchful eye. An index to no longer extant letterbooks indicates that in September 1853 there was some question as to the need for three lights at Nauset. Since that number of lighthouses remained at the site, the board obviously decided to retain them.[18]

Because they were deemed to have a lesser significance, the Nauset lights were given a lower priority for conversion to the Fresnel lens system. The letterbook index indicates that there was evidently some problem in placing the new lenses in the old lanterns when they arrived for installation. At any rate, the old lanterns were replaced in October 1858, and sixth order Fresnel lenses were placed in them. That lens was the smallest of its type. Valve lamps were used in conjunction with the lenses.[19] Figures 1 and 2 show the "three sisters" masonry towers with the 1858 lanterns. The exterior ladders were added to the towers in 1889. Also visible in the photograph are the masonry house and the barn.

---

17. Johnson, The Modern Light-House Service, pp. 20-21; Holland, America's Lighthouses, pp. 33-35. First order Fresnel lenses were the largest with a height of about 12 feet while the smaller ones - fourth, fifth, and sixth - were no more than 3 feet tall. The larger size allowed a bigger lamp to be installed and thus a greater brightness occurred

18. Index to Light-House Board Letterbooks, Records of the United States Coast Guard, Record Group 26, National Archives, Washington, D.C.

19. Ibid.

Several changes occurred to the lights in the post-1858 period. In 1869 the valve lamps were replaced with Franklin lamps. The sixth order Fresnel lenses were replaced by those of the fourth order in 1873 in each lighthouse lantern. By the early 1880s, Haines lamps with 1-5/8-inch burners were in use.[20] All of these lamps had concentric wicks. The difference came from the method used to deliver fuel to the wicks.

The Wooden "Three Sisters"

The Lighthouse Board annual report for 1892 stated, "three wooden movable towers were erected by hired labor, 30 feet to the westward of the old towers. This was done because of the washing away of the bank." Despite the jerry-built construction methods, the old masonry towers served for 54 years. Rather than deteriorating to disuse, they were claimed by the sea when the bank eroded. Part of their remains appear in the surf at certain low tides. The lanterns from the brick towers were moved to the new lights on April 25, 1892.[21]

The circular, wooden towers were taller than the ones they replaced. Raising the lights could perhaps have stemmed from the findings of the 1851 investigation which concluded that many lighthouses were too short to serve effectively. At any rate, the new lights were 22 feet to the top of the tower, with an additional 7 feet to the lantern's ventilator ball (Figure 3 and 4).

---

20. Annual Report of the Light-House Board to the Secretary of the Treasury for the Fiscal Year, 1869 (Washington, D.C.: Government Printing Office, 1869), p. 16; Annual Report of the Light-House Board of the United States, 1873 (Washington, D.C.: Government Printing Office, 1873), p. 18; Annual Report of the Light-House Board to the Secretary of the Treasury for the Fiscal Year Ending June 30, 1882 (Washington, D.C.: Government Printing Office, 1882), p. 17; Index to Light-House Board Letterbooks.

21. Annual Report of the Light-House Board to the Secretary of the Treasury for the Fiscal Year Ended June 30, 1892 (Washington, D.C.: Government Printing Office, 1892), p. 55.

Figure 1

The Original Masonry "Three Sisters"
Courtesy of the National Archives, Washington, D.C.
View from northwest, 1869-1875

The house at the right is the 1838 brick keeper's residence. The 1875 frame residence has not been built. The wing to the west of the barn may be the 9' by 15' addition built in 1869.

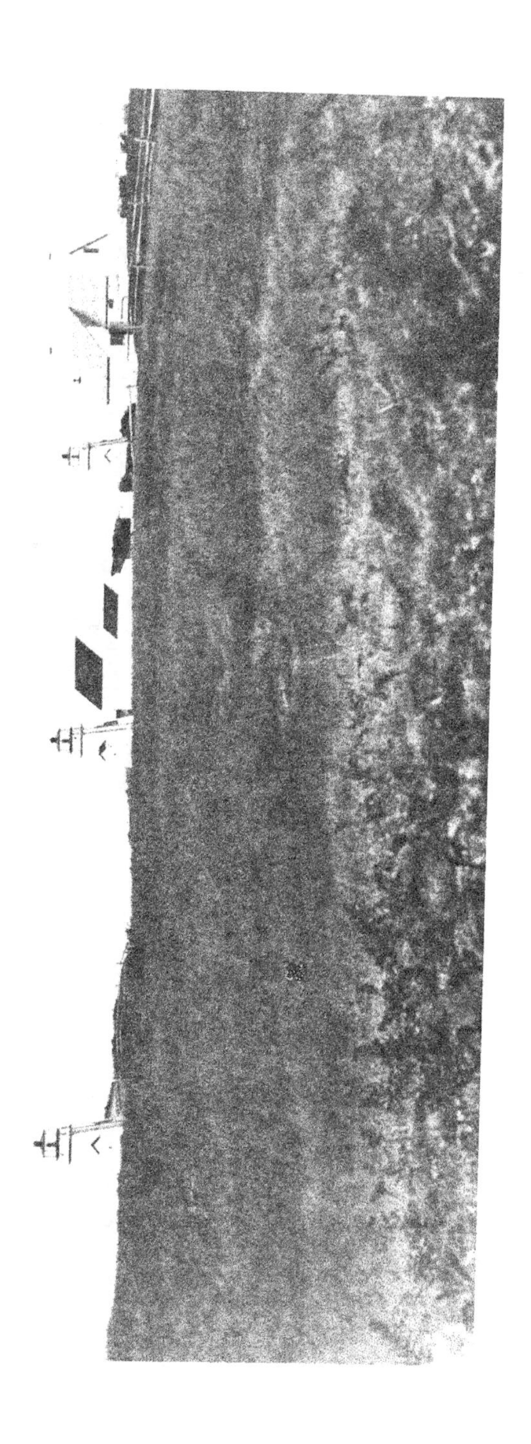

Figure 2

The Original Masonry "Three Sisters"
Courtesy of the National Archives, Washington, D.C.
View from southeast, 1875-1892

In the background one can see the 1838 brick house and the 1875 wood frame dwelling. The use of the term "three sisters" was never an official name for the lighthouse. According to Edward Rowe Snow in his book, Famous New England Lighthouses, they were named the "three sisters of Nauset" soon after their construction, but he does not indicate whether the local population or mariners gave them that designation.

Figure 3

The 1892 Wooden "Three Sisters"
Courtesy of Cape Cod National Seashore
View from southeast, 1895-1911

The storm doors attached to the entrances in 1895 are visible on two of the lighthouses.

Figure 4

The 1892 Wooden "Three Sisters"
Courtesy of the National Archives, Washington, D.C.
View from north, 1892-1921

The building in the background between the frame keeper's residence and the barn appears to be the French Cable Station which burned May 3, 1901.* The Cable Station building was sold and converted to use as a hotel in 1893 after the French Cable Company operations were moved to Orleans in 1891. Because of the photographer's viewpoint, the building cannot be the Cable Company's stable which is shown on map 2.

*Interview of Alice Snow by Berle Clemensen, June 22, 1983, confirmed in telephone conversation February 20, 1935.

The foundations were formed from posts that had been driven 4 feet into the ground. In 1910 each tower was described as having a base diameter of 14 feet, with 10-1/2 inch thick walls. At the top, the diameter diminished to 11 feet 2 inches with 8 inch thick walls Architectural investigation, however, has shown the walls to be 6 inches thick. A wooden deck with a corbelled cornice extended beyond the tower walls. White-painted shingles covered the tower exteriors while the inside walls were lath and plaster. Each tower had one window and one door. The double sash window contained 6 over 6 glazing in each sash. In addition the sashes measured 28 inches wide and 26 inches high. Inside the towers, a wooden stairway led to a single landing that covered half of each structure's diameter. A metal ladder from the landing to a trap door gave entrance to the lantern.[22]

The old nine-sided lanterns from the brick lighthouses topped the towers. Wooden parapets comprised the lower portion of the lanterns while the remainder, except for the copper roof and ventilation balls, was cast iron. Each lantern contained nine plates of glass that measured 3 feet high and 28 inches wide. The parapet, below the glass, had ventilators and a wooden door which led to the iron balustraded deck.[23]

The fourth order Fresnel lenses, placed in the brick lighthouse lanterns in 1873, were transferred with the lanterns to the wooden towers. Both the north and center tower lenses were manufactured in France by Barbier and Fenester, while the one in the south tower, also imported from that country, was made by Henry Lepaute. At some time during this period, improved Funck Heap lamps with one wick and burner were installed in the lenses.[24]

---

22. "Description of Light-House Towers, Buildings, and Premises at Nauset Beach, Massachusetts, Light-Station, January 11, 1910," pp. 2-3, First District Coast Guard Files, Boston, Massachusetts.

23. Ibid., pp. 3-4.

24. Ibid., pp. 4-5.

Only one change was made to the Nauset lights between 1892 and 1911. In 1895 a storm porch was placed at the door of each tower.[25]

By 1911, continuing erosion brought the Nauset Beach cliff to within 8 yards of the north tower and 18 yards of the one farthest south. As a result, it was decided to move the central light back from the cliff and use it as the only lighthouse into which a flashing light would be placed. Although technology had advanced to the point that a new, single flashing light could produce 20 times the brilliance as compared to that of the three lights combined, illuminating power was not the only reason to reduce the lighthouse number to one. Three lights could be mistaken for one or two if a vessel sailed by them at certain angles. In addition, the double and triple light systems in separate structures were being phased out for economic reasons.[26]

The central tower of the "three sisters" was moved near the 1875 wood frame keeper's quarters and attached to it by a short covered walkway. It rested on a solid brick foundation (Figure 5). A small hole was cut near the center of the deck to allow the installation of part of the revolving apparatus in the interior below the lantern. A fourth order Fresnel lens with a ball bearing flashing apparatus was installed. It included the Funck Heap oil-wick lamp. Soon, however, an incandescent

---

25. Annual Report of the Light-House Board to the Secretary of the Treasury for the Fiscal Year Ended June 30, 1895 (Washington, D.C.: Government Printing Office, 1895), p. 54.

26. Inspector H. C. Poundstone, Second District, to the Commissioner of Lighthouses, Washington, D.C., May 31, 1911, Correspondence of the Bureau of Lighthouses 1911-39, File 1342-E, Box 953, Records of the United States Coast Guard, Record Group 26, National Archives, Washington, D.C.; Commissioner of Lighthouses to the American Association of Masters, Mates, and Pilots, Piscataqua Harbor No. 83, January 24, 1912, Correspondence of the Bureau of Lighthouses 1911-39, File 1342-E, Box 953, Records of the United States Coast Guard, Record Group 26, National Archives, Washington, D.C.

oil-vapor lamp was installed. To retain a touch of the past, the light produced a triple flash every 10 seconds when it went into operation on June 1, 1911.[27]

Within days after the single light began to function, the second district inspector reported that the tower was so poorly constructed that, despite its low height, even moderate winds caused it to vibrate. The shaking interfered with the flashing apparatus. The inspector asked permission to install four galvanized guy wires as a means of stabilizing the tower. The guy wires are visible in Figure 5.[28]

Inspector H.C. Poundstone recommended that the two unused towers be torn down, saving the lanterns and using the rest for the keeper's firewood. Before demolition approval was received, H.N. Cummings of North Eastham appeared in Poundstone's office and asked to buy the two towers. As a result, he contacted the commissioner of lighthouses and requested permission to sell them--minus the lanterns--by sealed bid. A month later, E.L. Hoyt of Chicopee Falls, Massachusetts, wrote to the

---

27. Annual Report of the Commissioner of Lighthouses to the Secretary of Commerce and Labor for the Fiscal Year Ended June 30, 1911 (Washington, D.C.: Government Printing Office, 1911), p. 40; Inspector H. C. Poundstone, Second District to the Commissioner of Lighthouses, Washington, D. C., June 12, 1911, Correspondence of the Bureau of Lighthouses 1911-39, Files 1342-E, Box 953, Records of the United States Coast Guard, Record Group 26, National Archives, Washington, D. C.; Commissioner of Lighthouses to the American Association of Masters, Mates, and Pilots, Piscataqua Harbor No. 83, January 24, 1912; Annual Report of the Commissioner of Lighthouses to the Secretary of Commerce and Labor for the Fiscal Year Ended June 30, 1912 (Washington, D. C.: Government Printing Office, 1913), pp. 43, 47; Inspector, Second District to the Commissioner of Lighthouses, Washington, D. C., September 30, 1912, Correspondence of the Bureau of Lighthouses 1911-39, File 1342-E, Box 953, Records of the United States Coast Guard, Record Group 26, National Archives, Washington, D.C.

28. Inspector H.C. Poundstone, Second District to the Commissioner of Lighthouses, Washington, D.C., June 12, 1911.

commissioner about purchasing the towers. He was told that they would be sold at public auction.[29] For some unknown reason, however, the sale was delayed.

Seven years later in 1918, the two lighthouses were sold. Helen M. Cummings, whose husband had previously inquired about the towers, bought the structures at a three-bid auction. She paid $3.50 for both buildings and had to remove them from the property within 10 days. They were removed within the allotted time and placed on a site near the French Cable Hut where they remained for two years. During that period, the Cummingses used one tower for a kitchen and living room on the ground level, and they used the landing as a bedroom. The other tower contained two bedrooms. In 1920 the two structures were moved to their present location where they were butted into a 26- by 28-foot living room. A morning room and kitchen were also added.[30]

In addition to attaching the two towers to a central structure, the Cummingses made some changes to the buildings. A second ground-floor window was placed in each and another window added at the second-floor landing level of both structures. The landing in each tower was enlarged from covering half the space to that of about three-quarters of the area. In conjunction with that alteration, the wooden stairs leading to each landing were removed and relocated to another position. A nearly flat roof covered with tar paper replaced the old deck covering. (Figure 7)

---

29. Inspector H.C. Poundstone, Second District to the Commissioner of Lighthouses, Washington, D.C., May 31, 1911; Inspector H.C. Poundstone, Second District to the Commissioner of Lighthouses, Washington, D.C., June 26, 1911, Correspondence of the Bureau of Lighthouses 1911-39, File 1342-E, Box 953, Records of the United States Coast Guard, Record Group 26, National Archives, Washington, D.C.; Commissioner of Lighthouses, Washington, D.C. to E. L. Hoyt, July 26, 1911, Correspondence of the Bureau of Lighthouses 1911-39, File 1342-E, Box 953, Records of the United States Coast Guard, Record Group 26, National Archives, Washington, D.C

30. John A. Cummings written account of the sale of two of the Three Sisters to his parents, March 23, 1969, Cape Cod National Seashore park files.

Figure 5

The Wooden Central Tower Attached to the 1875
Wood Frame Keeper's Dwelling
Courtesy of the National Archives, Washington, D.C.
View from northeast

This May 1915 photograph shows the original tower window which was identical to the ones in the other two lights. Guy wires placed on the tower in 1911 to stabilize it are also visible as well as the connection to the 1875 wood frame keeper's dwelling.

Figure 6

Installation of the Cast-Iron, Unused Chatham Tower
Courtesy of the National Archives, Washington, D.C.
View from southwest, 1923

In addition to the cast-iron tower, the 1875 wood frame keeper's dwelling can be seen on its new location over a new cellar. This photograph of the 1875 structure was taken on the opposite side of the house from that shown in Figure 5 and, therefore, accounts for the different appearance.

Figure 7

"The Twins" and attached cottage
Courtesy of Cape Cod National Seashore
View from southeast, 1976
Photograph by Swan

The Cummings' cottage incorporated the North Twin (right) and the South Twin (left).

Figure 8

The Beacon Cottage
Courtesy of Cape Cod National Seashore
View from northeast, July 1973
Photograph by Swan

The Hall family converted the Beacon into a cottage by adding wings to the west and south.

Figure 9

Postcard Views of Three Sisters
Courtesy of the Society for the Preservation of New England Antiquities

This postcard shows separate views of each of the Three Sisters. The caption "donated 1919" appears on a piece of paper glued to the back of the card. The fact that all three views are printed on one card strongly suggests that they were taken at the same time. The brackets under the Center Beacon cornice indicate that the photograph was taken after June 12, 1911. Note that the lanterns of the north and south beacons retain their lanterns but are either missing their shades or do not have them drawn. This fact suggests that the lighting apparatus had been removed because sunlight caused the lenses to darken.

Nauset Beach Lights,
The North Beacon, Eastham, Mass.

Nauset Beach Lights,
The South Beacon, Eastham, Mass.

Nauset Beach Lights,
The Centre Beacon, Eastham, Mass.

In December 1965 the National Park Service obtained the two towers from the Cummingses' son, John. Since that time, in 1980, the additions have been removed, leaving the two structures freestanding.

The third of the "three sisters" remained in service until 1923. At that time the disused cast-iron north tower at Chatham was moved to Nauset Light Beach and replaced the wooden third sister (Figure 6). Presumably, the fourth order lens was transferred to the metal tower at the time and is the one now on display in the Salt Pond visitor center. In 1923, when the Chatham tower was moved to Nauset Light Beach, the 1875 keeper's dwelling was moved to a new location (Figure 6). A concrete foundation was laid for the cast-iron tower. In that same year, Albert Hall purchased the wooden tower with lantern and moved it to its present location.[31] He built a single-story addition to the tower and named it the "Beacon." Hall's modifications were similar to those made by the Cummingses. Windows were added at the ground-floor and landing levels. The landing area was enlarged, but the wooden stairs remained in their original position. (Figure 8)

Albert Hall's son, Harold, sold the structure to the National Park Service. In the summer of 1983 the one-story addition was removed, leaving the tower freestanding, and on September 13, 1983, it was moved to a site adjacent to the other towers.

Ancillary Structures

As shown on the 1840s map, only two other buildings were constructed in 1838 when the lighthouses were erected. These structures included a brick, one-story keeper's dwelling and a privy. The house measured 34 feet by 20 feet, with a 16- by 14-foot kitchen attached to the rear.

---

31. History of "Jack O' Lantern"--Hall Cottages, a telephone conversation between Doris Doane (a Cape Cod National Seashore employee) and Harold T. Hall, November 21, 1975, Cape Cod National Seashore files.

The main portion contained two rooms with a central stair leading to the attic. A 4- by 5-foot privy was north of the house.[32]

At an unknown date (perhaps the 1850s), a board and batten barn was constructed on a site west of the north tower. Its location was recorded on the 1885 and 1922 maps. The first mention of the barn was in the 1868 annual report when it was reported to have been repaired. A 9- by 15-foot addition was attached to the rear of the barn in the following year.[33]

The 1869 annual report indicated that a 9-1/4- by 18-1/2-foot shed was built at the Nauset Beach light site. It could have been the hen house that appeared on the 1885 map near the barn.[34]

In 1873, Nauset Beach received an assistant keeper because of the additional work required by the change in lenses from sixth order to fourth order. The assistant keeper lived in the same dwelling as the keeper, an arrangement that did not prove satisfactory. As a result in 1875, a new wood frame, two-story keeper's dwelling was constructed. The assistant keeper was assigned the 1838 brick structure for his house. The new quarters were about 95 feet west of the middle tower and just north of the old dwelling.[35]

---

32. Ground Plan Map of Nauset Beach Structures, no date, United States Coast Guard Academy Library, New London, Connecticut; report of I.W.P. Lewis, p. 34.

33. Annual Report of the Light-House Board to the Secretary of the Treasury, for the Year 1868 (Washington, D.C.: Government Printing Office, 1868), p. 16; Annual Report of the Light-House Board, 1869, p. 16.

34. Annual Report of the Light-House Board, 1869, p. 16.

35. Annual Report of the Light-House Board of the United States to the Secretary of the Treasury for the Fiscal Year Ending June 30, 1874 (Washington, D.C.: Government Printing Office, 1874), p. 21; Annual Report of the Light-House Board to the Secretary of the Treasury for the Fiscal Year Ending June 30, 1876 (Washington D.C.: Government Printing Office, 1877), p. 16.

Map 1

Early Site Plan (ca. 1840s?) of Nauset Beach Light Station

The original longhand caption reads:

Description of Land for Nauset (sic) Beach Light House . . . Beginning on land of Benjamin H.A. Collins, at a stake and stones, thence East, one degree South, thirty-one Rods to the Back side Bank, thence North, one degree East, by the Bank, twenty-six Rods to a stake and stones on said Bank, thence West one degree North, thirty-one Rods to a stake and stones, thence South one degree West twenty six Rods to the first bounds specified
Containing five acres and Six rods D.B.

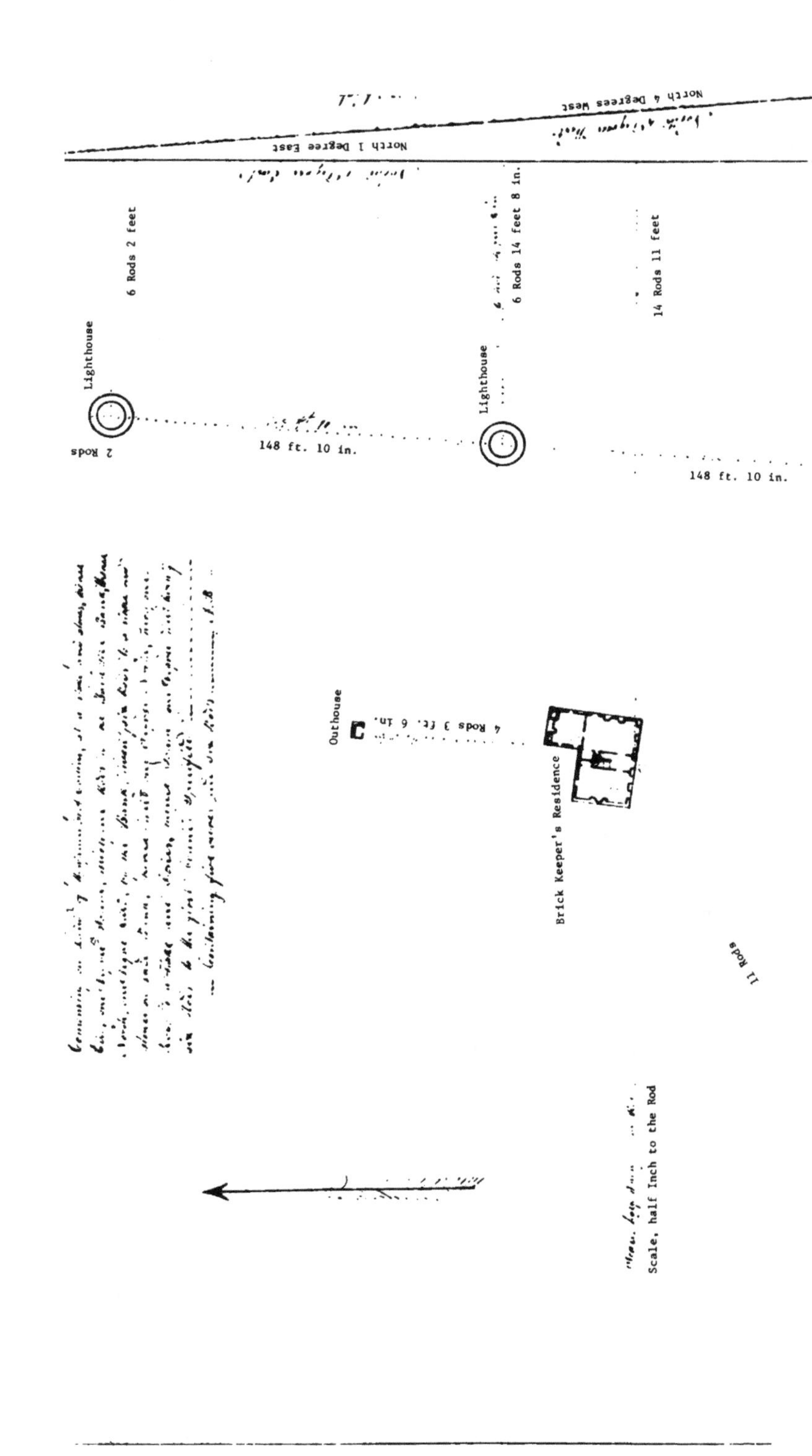
North 4 Degrees West
North 1 Degree East
6 Rods 2 feet
6 Rods 14 feet 8 in.
14 Rods 11 feet
Lighthouse
Lighthouse
2 Rods
148 ft. 10 in.
148 ft. 10 in.
Outhouse
4 Rods 3 ft. 6 in.
Brick Keeper's Residence
11 Rods
Scale, half Inch to the Rod

MAP 1

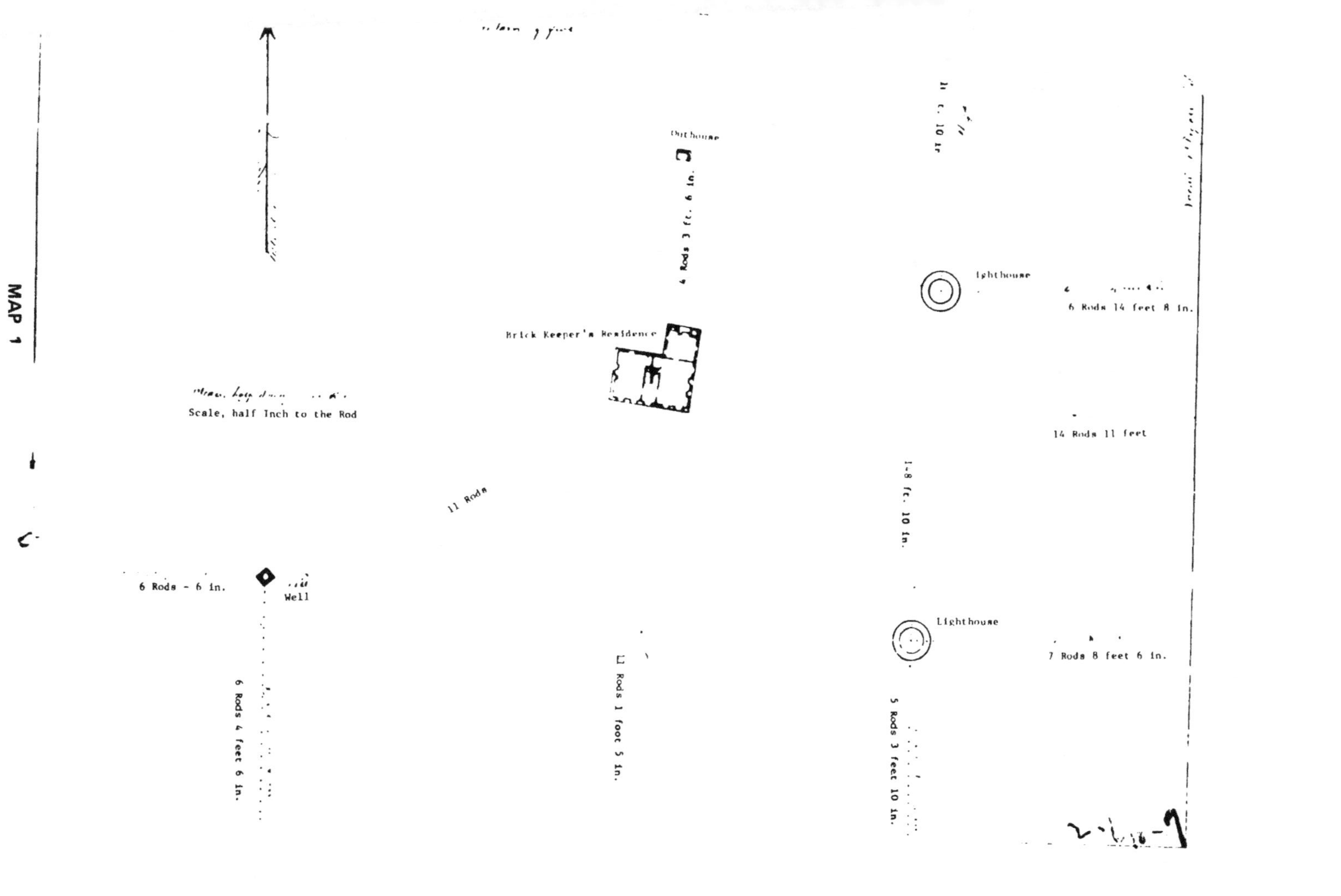
Outhouse
4 Rods 3 ft. 6 in.
Lighthouse
6 Rods 14 feet 8 in.
Brick Keeper's Residence
Scale, half Inch to the Rod
14 Rods 11 feet
1-8 ft. 10 in.
11 Rods
6 Rods - 6 in.
Well
Lighthouse
7 Rods 8 feet 6 in.
6 Rods 4 feet 6 in.
11 Rods 1 foot 5 in.
5 Rods 3 feet 10 in.

MAP 1

NAUSET BEACH, MASS.

LIGHT STATION.

Three towers and lights at Eastham on the E. side of Cape Cod, Mass.

Lat. 41° 51' 37" N.
Long. 69° 57' 4" W.

Reservation Surveyed Dec. 24 to 31, 1885, by E.P. ADAMS, C.E., L.H. Surveyor.

Scale, 1080

Base of Towers above water level 75 ft. *When rebuilt or renovated Refitted 1856, new dwelling 75.*
Focal Plane " " " 93 ft. *Area of Reservation within boundaries* 5.037 Acres
Direction of line of Towers, N. 4° 30' W. & S. 4° 30' E. *Area enclosed to edge of cliff* 4.80 "

NOTE. - The Magnetic Declination in 1837, given below, was estimated in 1885, by comparison of authorities; the declination at time of survey was determined by observation of Polaris.

— EXPLANATIONS. —

Mag. Dec. in 1837 est. 9° 40'; det. at survey 12° 30' W. Courses given on Plan are True bearings.

Light Towers. A. Old Dwelling. B. New Dwelling. C. Privy and Hen house.
D. Barn. E. Well house. F. French Cable Co's Stable. G. Gateway.
Boundary of Reservation. Line of Fence. • Stake, N & S line.
M.H.W. Line. M.L.W. Line. ▪ Stone Bound, 3 ft long 6"-6" top cut thus.

Feet.
100 0 100 200

E. P. Adams
L.H. Surveyor 1st & 2d Dists.

U.S.
L.H.E.

Wm. Stanton, Maj. of Engrs.
Engr. 1st & 2d L.H. Dists.

OFFICE OF L.H. ENGR. P.O. BLDG. BOSTON, MASS., MAR. 27, 1886

MAP 2

NAUSET BEACH, MASS.

LIGHT STATION.

Three towers and lights at Eastham on the E. side of Cape Cod, Mass.

Lat. 41° 51′ 37″ N.
Long. 69° 57′ 4″ W.

Reservation Surveyed Dec. 24 to 31, 1885, by E. P. ADAMS, C.E., L.H. Surveyor.

Scale, 1080

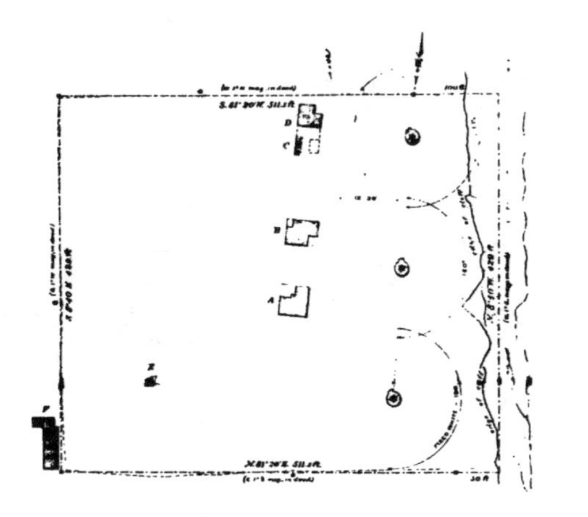

Kind of Lights	Sea Coast	Site Purchased	Sept 14 1837
Order of Lights	Three of 4th.	Deed Recorded in Barnstable Co. Reg. Bk 16 p 261.	
Characteristic of Light	Fixed White	First Buildings when built	1837
Base of Towers above water level	75 ft.	When rebuilt or renovated Re-fitted 1856, new dwelling 75	
Focal Plane	93 ft	Area of Reservation within boundaries	5.037 Acres
Direction of line of Towers, N. 4° 30′ W & S. 4° 30′ E.		Area enclosed to edge of cliff	4.00

NOTE.- The Magnetic Declination in 1837 given below, was estimated in 1885 by comparison of authorities; the declination at time of survey was determined by observation on Polaris

— EXPLANATIONS —

Mag. Dec. in 1837 est. 9° 40′; dec at survey 12° 30′ W. Courses given on Plan are True bearings

● Light Towers A Old Dwelling B. New Dwelling C Privy and Henhouse

D. Barn E Well house F French Cable Co.'s Stable G Gateway

— — — Boundary of Reservation - - - - Line of Fence • Stake, N°4 line

M. H. W. Line — M. L. W. Line ▫ Stone Bound 3 ft long 6″-6″ top cut thus — U.S. L.H.R.

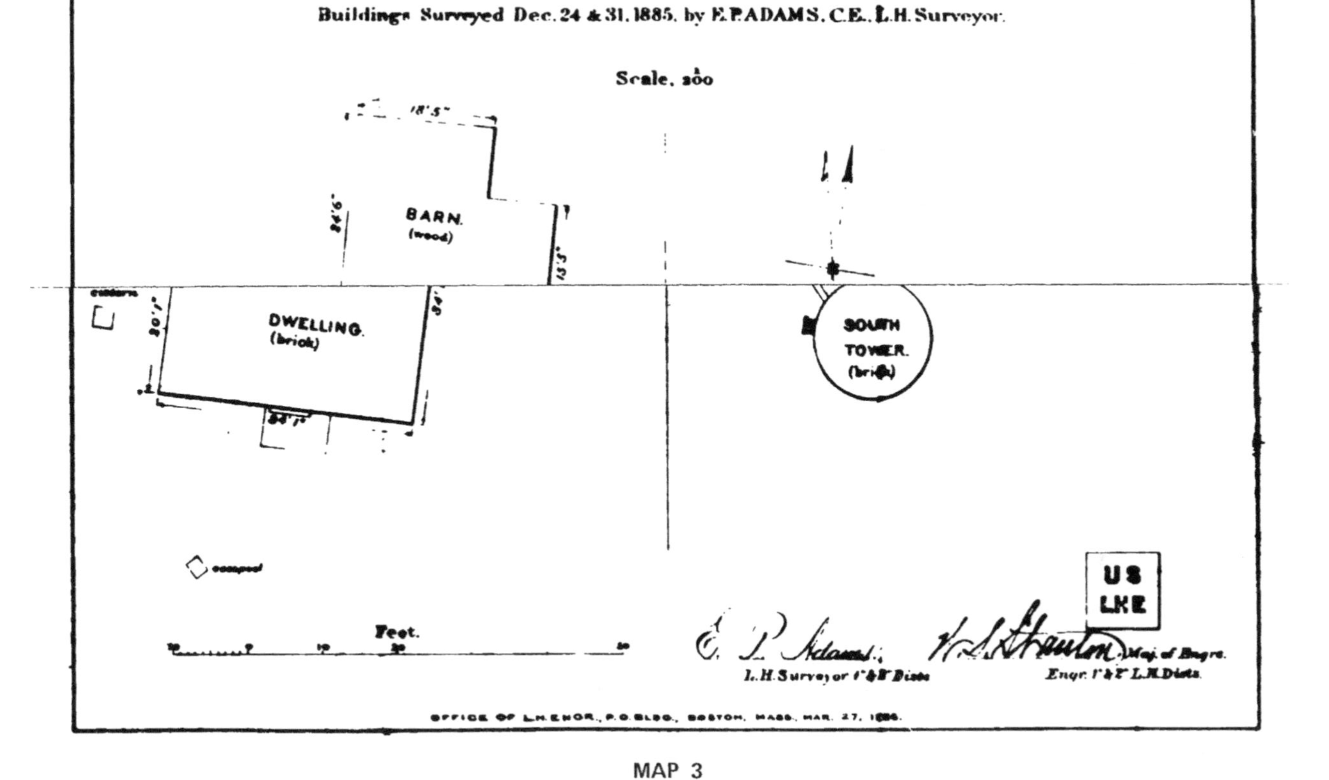
NAUSET BEACH. MASS.
LIGHT STATION.
Buildings Surveyed Dec. 24 & 31, 1885, by E.P. ADAMS, C.E., L.H. Surveyor.
Scale, 1/200
18'5"
BARN.
(wood)
24'6"
13'3"
DWELLING.
(brick)
SOUTH
TOWER.
Feet.
L.H. Surveyor 1st & 2nd Dists.
Maj. of Engrs.
Engr. 1st & 2nd L.H. Dists.
US
LHE
OFFICE OF L.H. ENGR., P.O. BLDG., BOSTON, MASS., MAR. 27, 1886.

MAP 3

NAUSET BEACH, MASS.

LIGHT STATION.

Buildings Surveyed Dec. 24 & 31, 1885, by E. P. ADAMS, C.E., L.H. Surveyor.

Scale, 200

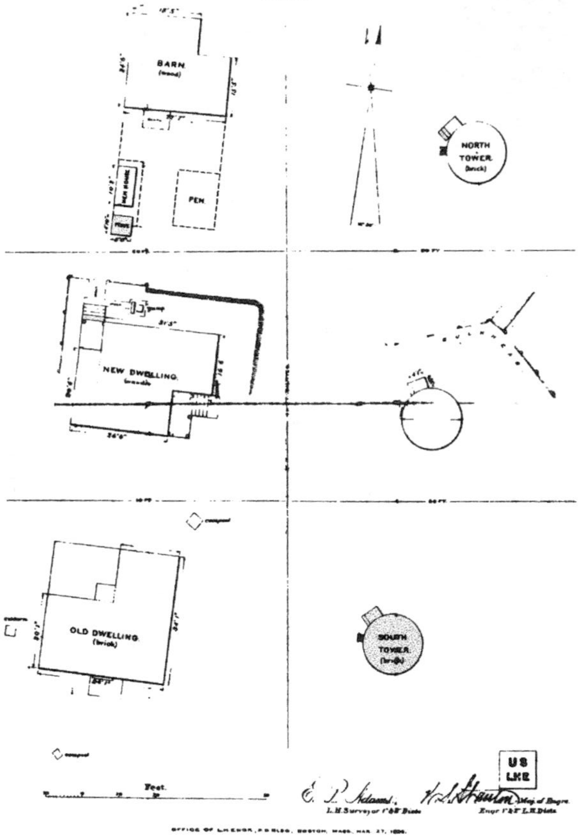

BARN
(wood)

PEN

HEN HOUSE

curb

NORTH
TOWER
(brick)

pump

NEW DWELLING
(wood)

compost

OLD DWELLING
(brick)

cistern

SOUTH
TOWER
(brick)

compost

US
LHE

Feet

E. P. Adams, L.H. Surveyor 1st & 2nd Dists.

Maj. of Engrs. Engr 1st & 2nd L.H. Dists.

OFFICE OF L.H. ENGR., P.O. BLDG., BOSTON, MASS., MAR. 27, 1886.

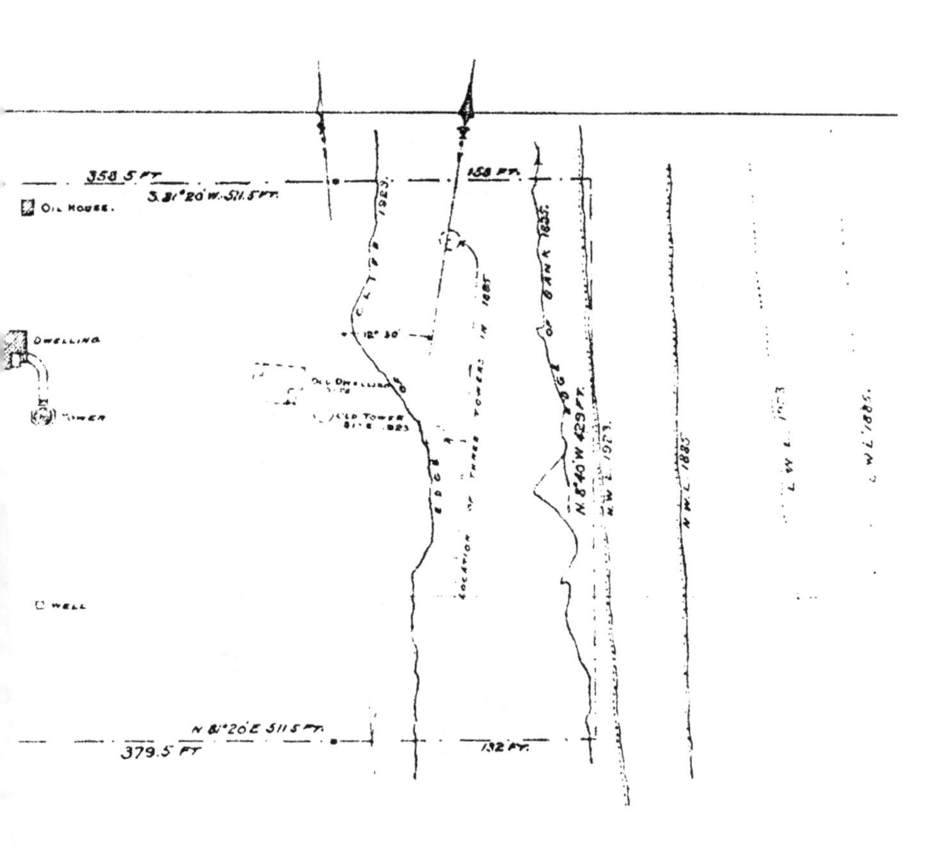

OFFICE OF LIGHTHOUSE SUPERINTENDENT,
SECOND DISTRICT, BOSTON, MASS.

NAUSET BEACH, MASS. LIGHT STATION. SCALE 1/1000

APPROVED FEB. 28, 1924.

FIRST ASST. SUPT. SUPERINTENDENT.

SURVEYED BY F.J.M.
CHECKED " "
TRACED " "

No. 539.

MAP 4

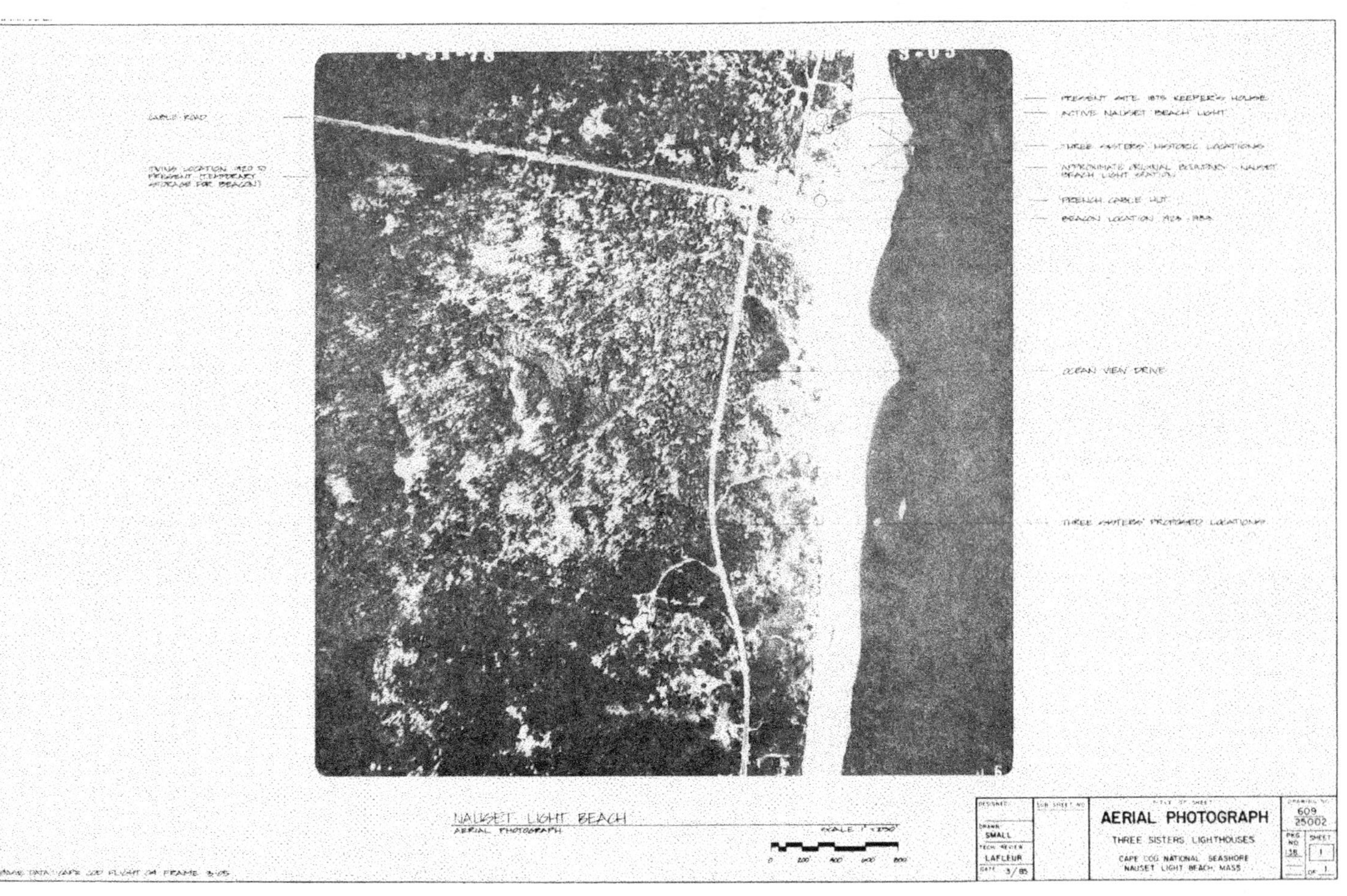
PRESENT SITE 1875 KEEPER'S HOUSE
ACTIVE NAUSET BEACH LIGHT
THREE SISTERS' HISTORIC LOCATIONS
FRENCH CABLE HUT
CABLE ROAD
OCEAN VIEW DRIVE
THREE SISTERS' PROPOSED LOCATIONS
NAUSET LIGHT BEACH
AERIAL PHOTOGRAPH
0 200 400 600 800
DRAWN SMALL
TECH REVIEW LAFLEUR
DATE 3/85
AERIAL PHOTOGRAPH
THREE SISTERS LIGHTHOUSES
CAPE COD NATIONAL SEASHORE
NAUSET LIGHT BEACH, MASS.
609
25002
PKG NO 38
SHEET 1 OF 1

MAP 5

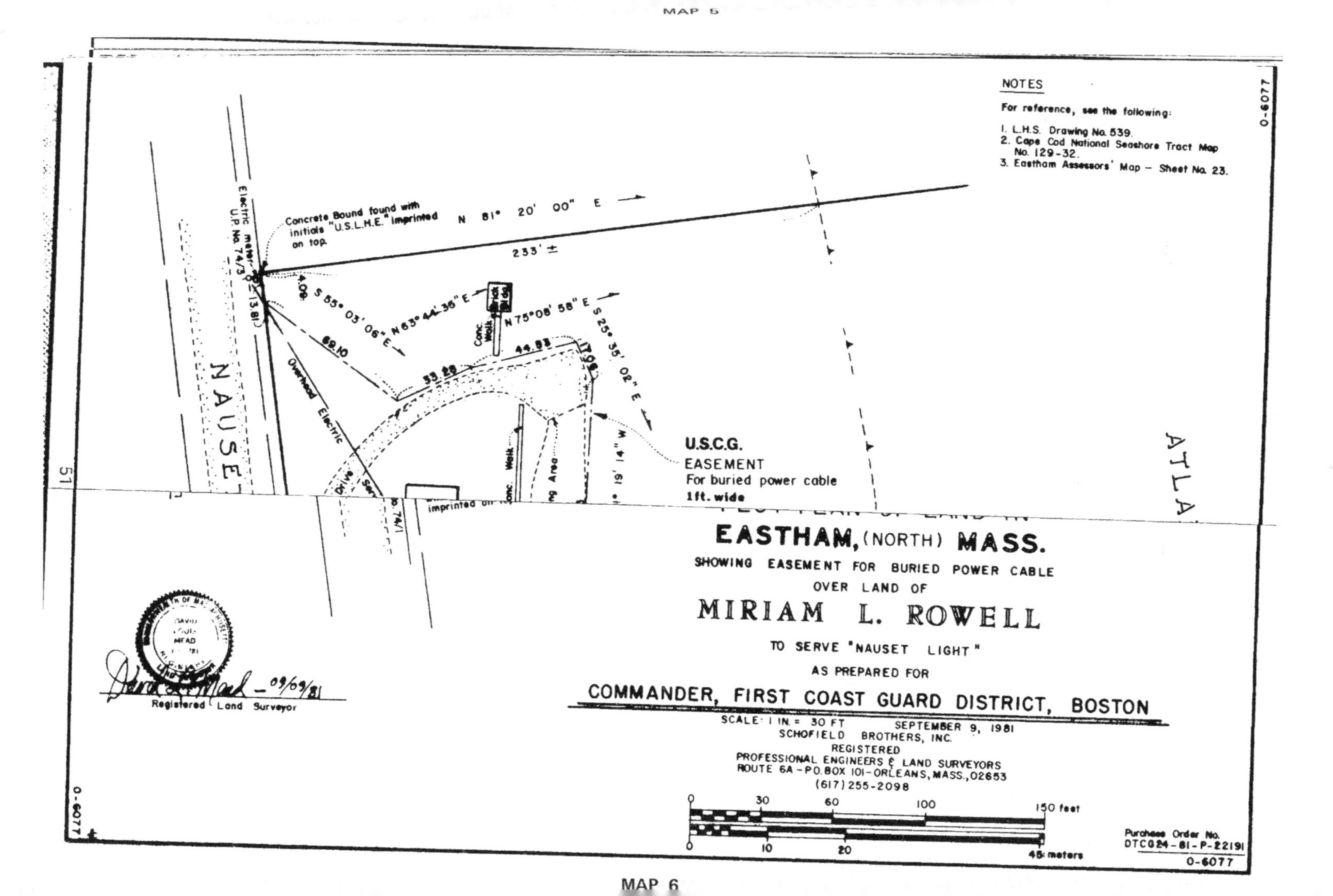

0-6077
NOTES
For reference, see the following:
1. L.H.S. Drawing No. 539.
2. Cape Cod National Seashore Tract Map No. 129-32.
3. Eastham Assessors' Map – Sheet No. 23.
Concrete Bound found with initials "U.S.L.H.E." imprinted on top.
N 81° 20' 00" E
233' ±
Electric meter U.P. No. 74/3
13.81
4.09
S 55° 03' 06" E
N 63° 44' 36" E
N 75° 08' 58" E
S 25° 35' 02" E
69.10
33.28
44.83
17.08
Overhead Electric
Conc. Walk
NAUSE
ATLA
U.S.C.G.
EASEMENT
For buried power cable
1ft. wide
EASTHAM, (NORTH) MASS.
SHOWING EASEMENT FOR BURIED POWER CABLE
OVER LAND OF
MIRIAM L. ROWELL
TO SERVE "NAUSET LIGHT"
AS PREPARED FOR
COMMANDER, FIRST COAST GUARD DISTRICT, BOSTON
SCALE: 1 IN. = 30 FT
SEPTEMBER 9, 1981
SCHOFIELD BROTHERS, INC.
REGISTERED
PROFESSIONAL ENGINEERS & LAND SURVEYORS
ROUTE 6A - P.O. BOX 101 - ORLEANS, MASS., 02653
(617) 255-2098
09/09/81
Registered Land Surveyor
0 30 60 100 150 feet
0 10 20 45 meters
Purchase Order No. DTCG24-81-P-22191
0-6077

MAP 6

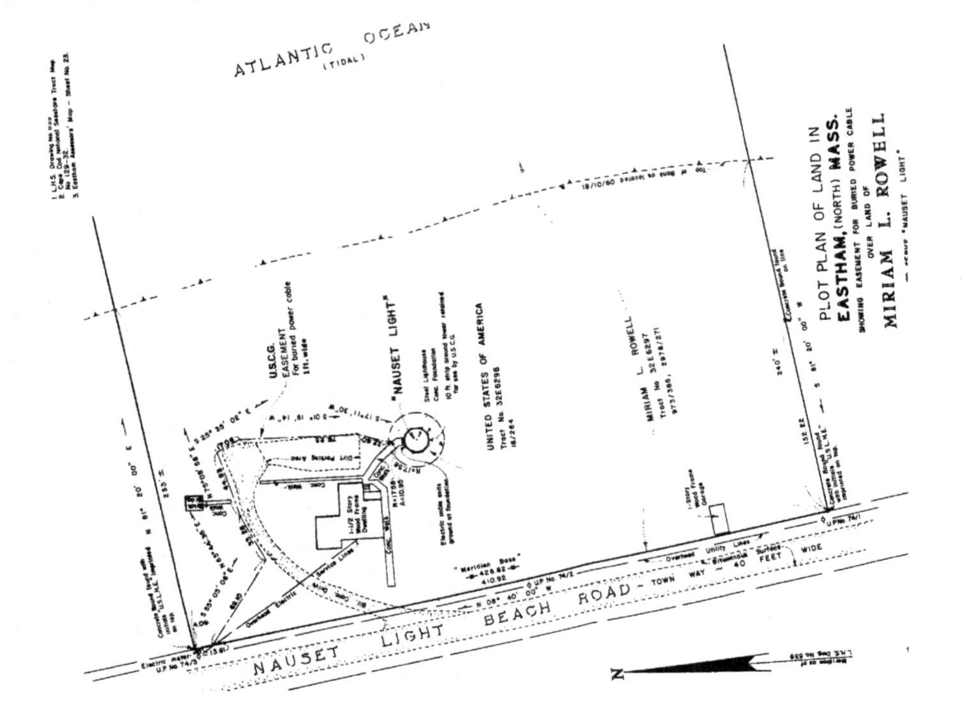
ATLANTIC OCEAN
(TIDAL)
PLOT PLAN OF LAND IN
EASTHAM, (NORTH) MASS.
SHOWING EASEMENT FOR BURIED POWER CABLE
OVER LAND OF
MIRIAM L. ROWELL
"NAUSET LIGHT"
U.S.C.G.
EASEMENT
For buried power cable
1ft. wide
UNITED STATES OF AMERICA
Tract No. 32E6298
MIRIAM L. ROWELL
Tract No. 32E6297
1-1/2 Story Wood Frame Dwelling
Dirt Parking Area
Steel Lighthouse
Conc. Foundation
"Meridian Base"
NAUSET LIGHT BEACH ROAD - TOWN WAY ~ 40 FEET WIDE
Overhead Utility Lines
Overhead Electric Service Lines
N

A brick oil house, 11 by 9-1/3 feet, was built in 1892 at the same time as the wooden towers. As shown on the 1922 map, it was approximately 90 feet west of the barn.[36]

At an unknown date, a white signal mast that was 65 feet high was erected on the site. A 1910 description of the Nauset Beach lights placed it 83 feet southeast of the middle tower.

The final structure to be built at the site was a 9- by 18-foot garage. Built between 1922 and 1927, it was 200 feet southwest of the cast-iron tower. It evidently signaled that the Lighthouse Service recognized the advent of the automobile age.[37]

After 1883, when an assistant keeper's services were no longer required at Nauset Beach, the old brick dwelling remained vacant for a time. Soon, however, it was used to store firewood and paint. After a recommendation, the house was demolished in late 1912.[38]

In 1923, at the time the unused cast-iron tower was moved from Chatham to Nauset, the 1875 wood frame keeper's quarters was moved back from the bluff's edge and installed over a new cellar.[39] It is currently on that site..

At some unknown date between 1922 and 1927, the barn, hen house, and privy were removed. At least the 1927 description of Nauset Beach light did not mention the existence of those buildings.

---

36. Annual Report of the Light-House Board, 1892, p. 55.

37. Nauset Beach Light Description, June 9, 1927, Description of Light Stations 1876-1938, Box 5, Records of the United States Coast Guard, Record Group 26, National Archives, Washington, D. C.

38. Inspector Goddard to the Commissioner of Lighthouses, October 4, 1912, Correspondence of the Bureau of Lighthouse 1911-39, File 1342-A, Records of the United States Coast Guard, Record Group 26, National Archives, Washington, D. C.

39. Annual Report of the Commissioner of Lighthouses to the Secretary of Commerce for the Fiscal Year Ended June 30, 1923 (Washington, D.C.: Government Printing Office, 1923), p. 48.

Supplies and Supply Methods

Prior to the advent of Stephen Pleasonton's administration in 1820, lighthouses were supplied by a number of local contractors. These individuals not only made repairs but furnished oil, wicks, chimneys, and cleaning stores. From the period 1812 to the early 1850s, Winslow Lewis was the sole source of lamps and reflectors. After Pleasonton, as fifth auditor of the treasury, gained oversight of the Lighthouse Establishment, he instituted a system by which one contractor supplied the needs for all lights and repaired them as well. The contractor was to visit each light station annually and report on its condition and the conduct of the keeper. As a check on the contractor, each keeper was to make a yearly report on the supplies he received.[40]

In 1840 the Lighthouse Establishment acquired its first tender. It was a sailing vessel that was transferred from the United States Revenue Cutter Service. The ship received limited use, however, since a contractor continued to service most lights. A supply vessel visited each light on an annual basis. The delivery of supplies began in April in Georgia. From there, the vessel worked its way north finishing in Maine in September. In October the ship would return to Florida and work its way to Louisiana. Supplies often arrived at Nauset in late July or early August. The first extant record of delivery to Nauset Beach occurred on July 31, 1850. The keeper, Joshua Crosby, received 100 tube glass, five gross of wicks, 47 yards of cloth, two buff skins, two boxes of tripoli (an abrasive cleaner), two boxes of soap, three pairs of scissors, two burners complete, one fountain (lamp fuel container), one hand lamp, and one file.[41]

Before 1860 all of the lighthouses were supplied with sperm oil from New Bedford, Massachusetts. As increasing industrialization occurred

---

40. Johnson, The Modern Light-House Service, p. 15.

41. Letter from the Secretary of the Treasury transmitting the Report of the General Superintendent of the Lighthouse Establishment, December 19, 1850, p. 63, Records of the United States Coast Guard, Record Group 26, National Archives, Washington, D.C.

with its dependence on sperm oil, the price soared. During the 1841-42 period, sperm oil sold for $.55 per gallon. By 1850-51 it had reached $1.17, and in 1855 it cost $2.25. The Lighthouse Service began to search for an alternative fuel. In the late 1850s colza or rapeseed oil seemed to answer the need, but inadequate supplies forced the Lighthouse Board to search for another substitute. Experimentation showed that lard oil was the solution to the problem.[42]

With the introduction of lard oil, the Lighthouse Board established its own distribution system. The old customs station at Tompkinsville on Staten Island, New York, was converted to an oil supply center. Although slowed in its procurement of vessels by the Civil War, the board acquired two schooners (the Pharos and the Guthrie) for use in making East Coast deliveries. These two ships carried lard oil in large casks. When one of the schooners arrived at a light station, it would lower a small boat filled with small barrels along side. The oil would be pumped into the small casks and rowed ashore. On a number of occasions the small boats overturned in heavy surf, dumping their contents into the sea.[43]

By the 1870s the Lighthouse Board had converted its Staten Island center into a warehouse nucleus for all supplies. In 1876 it replaced the two East Coast schooners with one vessel--the Fern. At this time the Lighthouse Service began to name its vessels after plants and flowers

In 1878, when conversion to kerosene began with fourth, fifth, and sixth order lights, the Fern had a carrying capacity of 30,000 gallons of that fuel. It was carried and delivered in small cases. The Fern usually arrived at Nauset Beach in early August to deliver that light station's supplies.[44]

---

42. "Light-House Construction and Illumination," Putnam's Magazine, 8 (August 1856), p. 208; Holland, America's Lighthouses, p. 23.

43. Kirk Munroe, "From Light to Light: A Cruise of the Armeria Supply Ship," Scribner's Magazine, 20 (October 1896), pp. 460-61.

44. Munroe, "From Light to Light," p. 461.

In 1891 the Fern was replaced by the Armeria (botanical name for Sweet William). It had a carrying capacity of 100,000 gallons of fuel that was loaded and delivered in 5-gallon tins, each in a wooden case. (See appendix A for illustrations and a description of the Armeria's sister ship, the Azalia. Such information was not available for the Armeria.) Brick oil houses were constructed at this time at each light station to accommodate a 13-month oil supply in 5-gallon tins.[45]

The Armeria brought supplies to Nauset Beach a little later each year than the Fern, coming in late August to early September. In addition to oil, it carried such items as paint, lubricating oil, cans of tallow, brooms, mats, shovels, hoes, rakes, lamps, boxes of chimneys, window glass, soap, and tin chests (known as supply boxes) that held linen, stationery, brushes, cleaning material, and other minor articles. The keeper exchanged each empty 5-gallon kerosene container for a full one and had to save his worn brooms, mops, tools, and other equipment each year to exchange for new items. All worn equipment was taken back to the supply ship and eventually dumped at sea. (In 1896 the Armeria supplied over 300 lights in New England alone.)[46]

In the latter part of the 19th century a secondary depot for supplies to the second district was established at Wood's Hole on Cape Cod. Articles kept on inventory were buoys and appendages, fuel, lubricating oil, anchors, light-vessel chains, paint, and miscellaneous items. About the same time a lens and lamp repair shop was located in Boston to serve the Second Lighthouse Service District. Both of the above operations ceased about 1912.

Lighthouse Keepers

Prior to 1852 the Lighthouse Establishment operated on a very loose arrangement with its keepers. Many received their positions through

---

45. Ibid., pp. 461, 466.

46. Ibid., pp. 464-65, 467. Drawings of lighthouse equipment and tools will be furnished independent of the report to Cape Cod National Seashore.

political appointment. Most were untrained in the operation of the illuminating equipment and received only minimal instructions. In fact there were no written instructions outlining the keeper's duties. When I.W.P. Lewis inspected the Maine, New Hampshire, and Massachusetts light stations in 1842, he noted the poor performance of many keepers. The best individuals, he found, were old sailors who were more aware of the necessity of good coastal illumination. Many keepers were lazy and neglected to trim the lamp wicks between 11 and 12 o'clock at night, thus causing the lights to dim and lose their effectiveness. Some lit their lamps, but gave so little attention that if a light were to extinguish during the night, it would not be relighted. Lewis felt that the average yearly salary of $350 was too low and, therefore, discouraged qualified men from seeking such positions. Another regulation that he disapproved was the right of keepers to be pilots. He found that some of these men, to augment their income, would leave their wife and children in charge of the lighthouse while they acted as a pilot.[47]

With the establishment of the Lighthouse Board in 1852, the situation began to improve. The board instituted a requirement that keepers had to be at least 18 years of age and be able to read. It considered a keeper's job as a full-time position. Almost immediately, the board developed a set of written instructions defining the keeper's duties and the proper operation of a lighthouse. Although updated on a regular basis, the first such set of instructions can be found in pages 112-75 of the Annual Report of the secretary of the treasury, dated December 18, 1852. The instructions which were issued for 1911 are found in the United States Bureau of Lighthouse Regulations for the United States Lighthouse Service, Washington, D.C. (see appendix B).

Instruction manuals covered a multitude of subjects. The first duty of a keeper was to light the lamps precisely at sunset. Lights were to be kept free of smoke, with the flame at its greatest attainable height during

---

47. Report of I.W.P. Lewis, p. 57.

the entire night. A flame was to be periodically measured during the night. Lights were to be extinguished at sunrise whereupon the keeper was charged with making them ready for the next evening by 10 o'clock in the morning. Other instructions included directions for cleaning and placing the lamp chimneys, cleaning the lamps, trimming the wicks, preserving and economically using the oil, filling the lamps, keeping the lanterns free of ice and snow, and using the various tools.[48]

Despite the almost constant complaints of low pay, keepers' salaries did rise somewhat. By 1852 the Nauset Beach keeper received $425 per year. Just before the Civil War, average yearly wages rose to $600 and remained there until the early twentieth century. In 1910 the "three sisters" keeper was paid $679 but received no heating fuel allowance for his dwelling.

Starting in June 1911, when the three lights were reduced to one, his remuneration was adjusted down to $517 plus a $50 heating fuel allowance.[49]

For about a 10-year period between 1873 and 1883, an assistant keeper was stationed at Nauset Beach because the lights were upgraded from sixth order Fresnel lenses to ones of the fourth order. The usual salary of an assistant keeper was half that of the keeper. As a result, the subordinate at the "three sisters" received $300 per year.

Communication and Technology

The lighthouses at Nauset Beach served as one link in the lighthouse system by which mariners were warned of potential dangers along the

---

48. Charles Nordhoff, "The Light-Houses of the United States," Harpers New Monthly Magazine 48 (March 1874), p. 470.

49. Letter from the Secretary of the Treasury Transmitting Estimates of Appropriations required for the Service of the Fiscal Year Ending June 30, 1854, House of Representatives, Ex. Doc. No. 2, 32nd Cong., 2nd Sess., December 10, 1852, p. 97.

coast, in harbors, and on lakes and rivers. There was no communication among the various keepers. Instead, correspondence was kept, at first, with the local collector of customs who acted as a superintendent of lights. After 1852, with the establishment of a Lighthouse Board, a series of districts were created. Within these administrative units authority became more diffuse. The collector of customs retained the duty of appointment of keepers and financial matters, while an inspector saw to the building, maintenance, and supply of lights. Within a short time, however, the collector of customs' duties were given to the inspectors who headed each district. They received the title of general superintendent for their district.[50]

Technologically, the Nauset Beach lighthouses advanced as new innovations appeared in lighting and were adopted by the Lighthouse Service. The first method of lighting was the lamp and near parabolic reflector invented by Winslow Lewis. With the conversion to sixth order Fresnel lenses in 1858, a valve lamp was used for illumination. Franklin lamps were installed in fiscal year 1869. They were retained when the lenses were changed to fourth order in 1873. During the 1880s and part of the 1890s, Haines lamps were used. By the late 1890s, Funck Heap lamps provided the light. Shortly after the three towers were reduced to one in 1911, an incandescent oil vapor lamp served as the illuminant. It operated on the same principle as the present-day Coleman lantern and was the last oil-fired lamp used by the Lighthouse Service. In the 1920s the Nauset Beach light was converted to electricity with a bulb that produced an intensity of 30,000 English candlepower from the lens. In June 1981 the Coast Guard installed an airplane beacon-type lamp.[51]

---

50. Holland, America's Lighthouses, p. 36.

51. Annual Report of the Lighthouse Board, 1869, p. 16; Annual Report of the Commissioner of Lighthouses, 1912, p. 47; Nauset Beach Light Description, June 9, 1927.

ARCHITECTURE COMPONENT

Brief History of Physical Changes

The information presented here was extracted from the foregoing History Component and from historic photographs

Appearance, Occupancy, and Use

The first lighthouses to occupy the Nauset Beach site were three circular brick towers 15 feet high with base diameters of 16 feet and top diameters of 9 feet. They supported soapstone lantern decks of 12 feet diameter and octagonal iron lanterns. The contract for their construction called for the brick walls to be 3 feet thick at the base and 20 inches at the top. They were built in 1838. See Figure 1 for a historic photograph of the brick Three Sisters (with later lanterns) and see the history section pages 9 to 11. The contract also called for the construction of a one-story brick house 34 by 20 feet with a 16 by 14 foot kitchen wing, a 4 by 5 foot wood frame outhouse, and a well.

Sometime later, possibly in the 1850s, a board and batten barn was built. The earliest documentary evidence refers to its being repaired in 1868.

In 1858, sixth order fixed-light fresnel lenses were installed in new lanterns atop the brick towers.* The lenses had been delivered earlier, but it was discovered that they would not fit in the existing lanterns and their installation was delayed until new lanterns could be installed as well. The present lantern on the Beacon is one of the 1858 lanterns (at least from the sill up) which was transferred to the wooden towers in 1892.

---

*See the History Component pages 14-15, and Holland, America's Lighthouses, pages 33-35 for a discussion of Fresnel lens sizes.

In 1869, the board and batten barn received a 9 by 15 foot addition, and a 9 by 18 foot shed was built. Dimensional similarities suggest that the hen house which appears on the 1885 site plan may have been the 1869 shed (see Map 2).

The lenses were changed to fourth order fixed lights in 1873. This change resulted in an assistant keeper being assigned to the station, and to share quarters with the keeper.

In 1875 a new wood frame keeper's house was erected and the 1838 brick house became the assistant keeper's house. The 1875 keeper's house is still at Nauset Beach, though in a new location. Its present and original locations are shown on Map 4.

Maps 2 and 3 show the Nauset Beach Light Station in 1885.

Exterior stairs were added to the brick towers in 1889. Since they are present in Figure 1, that photograph can be bracketed between 1889 and 1892.

In 1892 the original Three Sisters were threatened by ongoing erosion of the sand scarp. Three new wooden towers, designed to be moved, were built 30 feet to the west, and the upper parts of the lanterns were moved to the new towers and placed on circular wood parapets. The material of the 1838 parapets cannot be determined from the photographs. It is the 1892 lighthouses that are the focus of this report. A 9 by 11 foot brick oil house was built the same year.

Vestibules (called "storm porches" in the documents) were added to the towers in 1895.

The 1910 "Description of Light-House Tower . . ." states that there is no well at the site, but there is a 2500 gallon brick cistern. It also describes a signal mast, 65 feet tall, and 83 feet southeast of the middle tower. The focal planes of the lanterns are listed as 97 feet above mean high water. The oil house is described as a 9 foot-11 inch by 5 foot-2 inch brick structure, located first 150 feet west of the north tower, then 264 feet west of the tower. These statements require clarification. Sources at the Coast Guard informed us that the inspection books were often used for several years with changes in conditions recorded by striking out the former condition and writing in the new. The double entry regarding the oil house is probably indicative of the change from three towers to one that occurred in 1911, but the use of the book after 1910 is not recorded on its cover which is dated January 11, 1909, with 1909 struck over and 1910 written in. Further proof of this practice occurs in the description of the lighting apparatus where description of the fixed white lights is struck over and a description of the flashing light is written in. We know that the flashing light was put into operation on June 1, 1911. The report first describes the location of the keeper's dwelling as 95 feet west of the middle tower, and then as attached to the tower by a short covered way. This change also occurred in 1911. The report also states that the old dwelling (presumably the 1838 keeper's residence) was used as a wood house and paint storage area. This entry is struck over, probably as a result of the late 1912 demolition of the old house. The barn is listed in the report, but there is no mention of the hen house or privy.

In 1911 the north and south lights were extinguished and the lighthouses became surplus property to be used as sources of parts for the remaining lighthouse. The central lighthouse was moved back from the advancing scarp and attached to the 1875 keeper's house with a covered walkway. A flashing light was installed in its lantern. The mechanism was apparently sensitive to lateral movement or vibrations, for the Second District Inspector soon requested permission to install guy wires to stabilize the tower. The guy wires, the tension ring to which they were attached, and the brackets which held the tension ring can all be seen in Figure 3.

The north and south towers apparently remained in place until 1918 when they were sold to Helen M. Cummings for $3.50, and were moved to a site near the French Cable Hut. The lanterns were retained by the Coast Guard.

A picture postcard from the collection of the Society for the Preservation of New England Antiquities shows three different views of the Three Sisters individually (Figure 9). The view of the central tower shows the guy wires that were installed in 1911 and the other photographs show the lanterns in place on the north and south towers. It is logical to assume that the three photographs printed together would have been taken at the same time. Therefore we can conclude that the lanterns remained on the towers for some time after they ceased being used. A photograph dated July 14, 1923, shows the present iron tower under construction. In the foreground, dimly visible, is a ventilator ball from a lantern. It is not clear whether it is from the Chatham light (the one that is now in place on the active Nauset Beach light) or from the north or south tower. It cannot be the one from the central tower since that one is still in place on the Beacon. In any case we cannot document any period of time when the three towers were in place, but the north and south lanterns were not.

When the north and south towers were temporarily located near the French Cable Hut, the Cummings family used the first floor of one tower as a kitchen/living room and the second floor and both levels of the other tower as bedrooms. In 1920 the Cummings family moved the towers to the present site and incorporated them as corner turrets in a new summer cottage which they called "The Towers." At this new site the towers retained their original north and south relationship. The Cummings family added six-over-six windows to both towers at first and second floor levels and increased the areas of the second floors.

In 1923 the unused iron tower was moved from Chatham and placed on a concrete foundation about 100 feet farther back from the edge of the scarp. The 1875 keeper's residence was moved back about the same distance, but not attached to the tower. See Map 5 for a current site plan. Both structures are presently in these locations. The central tower of the Three Sisters was sold to Albert H. Hall with its lantern intact, and he moved it to the location it occupied until September 13, 1983. He added wings to the south and west of the tower and used it as a cottage he named "The Beacon". He also added area to the second floor and two-over-two windows at both the first and second floors.

Sometime between 1922 and 1927 a 9 by 18 foot garage was built 200 feet southwest of the iron tower. In the same time period, the barn, hen house, and privy were removed.

All three of the historic towers were acquired by the National Park Service in the 1960s. The later cottage portion of the Twins, or "The Towers" was removed by the National Park Service in 1980 or 1981. The wings of "The Beacon" were removed in the summer of 1983 and the tower was moved to the site of the Twins on September 13, 1983. And that brings this brief history of physical changes to the Three Sisters right up to the present.

Setting

The evolution of the setting for the Three Sisters is not documented to the same extent as the evolution of the structures. The main sources of information are the photographs of the site, and the "Description of Light-House Towers, Buildings and Premises at Nauset Beach, Massachusetts, Light Station, January 11, 1910." It describes the soil as "sandy" and states that a public road passes the station. The nearest railroad, post office and town were two miles away in North Eastham. A 2500 gallon brick cistern was provided to collect rain water for drinking. The supply was considered ample. At the time of the report, there was no well at the site. There was a white signal mast, 65 feet high, 83 feet southeast of the middle tower. It is present in some of the historic photographs.

A comparison of recent photographs with the historic photographs shows that vegetation at the site is now taller and more dense than it was historically. There is also more residential and resort development in the vicinity now.

CHRONOLOGY OF STRUCTURES AT NAUSET BEACH LIGHT

	Lighthouse Towers	Light-Related Structures	Residences	Outbuildings	Site Features
1838	Brick towers built		Brick keeper's house built	Privy built	Well dug
1858	6th order fresnel lenses and present lanterns installed				
1850s?				Board & batten barn built	
1868				Barn repaired	
1869				9 by 15 addition 9 by 18 shed built	
1873	4th order lenses installed				
1875			Frame keeper's house built		
1889	Exterior stairs added				
1892	Frame towers built	11 by 9 brick oil house built			
1895	Storm porches built				
1911	North and south towers out of service			2500 gal. brick cistern present	65 foot signal mast present
1912	Central tower moved back and attached to residence guy wires installed		1838 brick house demolished		
1918	North and south towers sold and moved near cable hut				
1920	North and south towers moved to present site				
1922-27			9 by 18 garage built barn, hen house, and privy removed		
1923	Beacon moved to present site Chatham tower reerected		Frame house moved to present site		

Description of Existing Conditions

Because of the many similarities between the three lighthouse towers, the following description applies to all three with specific differences identified.

Exterior Description

Location

The Beacon is located adjacent to the upper parking lot at Nauset Light Beach, about 250 feet southeast of the intersection of Cable Road and Ocean View Drive in Eastham, Massachusetts. The tract number is 32-6837 and the building number is E-241. It sits near the top of a small hill at an elevation of 58 feet. UTM coordinates are: Zone 19 E.420-980 N.4634-390. (On September 13, 1983, it was moved to a site adjacent to the Twins.)

The Twins (north and south) are located about 250 feet north of Cable Road and 1100 feet west of its intersection with Ocean View Drive, also in Eastham, Massachusetts. Both are located on tract number 31-6121. North Twin is building number E-2464A. South Twin is building number E-246B. UTM coordinates are: Zone 19 E.420-510 N.4634-410.

Orientation

The orientation of circular structures is best expressed in terms of the location and orientation of a planar feature such as a doorway. The original entrance to the Beacon is presently located on its west side and the plane of the doorway is in a line 6 degrees east of magnetic north. The door of the South Twin is on its northwest side and is in a line 60 degrees east of magnetic north. The door of the North Twin is on its southwest side and the plane of the doorway is in a line 48 degrees west of magnetic north. Orientation is shown graphically on the drawings.

	Lighthouse Towers	Light-Related Structures	Residences
1838	Brick towers built		Brick keeper's house built
1858	6th order fresnel lenses and present lanterns installed		
1850s?			
1868			
1869			
1873	4th order lenses installed		
1875			Frame keeper's house built
1889	Exterior stairs added		
1892	Frame towers built	11 by 9 brick oil house built	
1895	Storm porches built		
1911	North and south towers out of service		
1912	Central tower moved back and attached to residence guy wires installed		1838 brick house demolished
1918	North and south towers sold and moved near cable hut		
1920	North and south towers moved to present site		
1922-27			9 by 18 garage barn, hen house and privy remov
1923	Beacon moved to present site Chatham tower reerected		Frame house mo to present site

Description of Existing Conditions

Becaue of the many similarities between the three lighthouse towers, the followi description applies to all three with specific differences identified.

Exterior Dec iption

Locatio

The Be on is located adjacent to the upper parking lot at Nauset Light Beach about 250 feet southeast of the intersection of Cable Road and Ocean i w Drive in Eastham, Massachusetts. The tract number is 32-6837 and e building number is E-241. It sits near the top of a small hill at an el ation of 58 feet. UTM coordinates are: Zone 19 E.420-980 N.4634-390. (On September 13, 1983, it was moved to a site adjacent to the Twins.)

The Tw (north and south) are located about 250 feet north of Cable Road a 1100 feet west of its intersection with Ocean View Drive, also in East m, Massachusetts. Both are located on tract number 31-6121. Not Twin is building number E-2464A. South Twin is building num r E-246B. UTM coordinates are: Zone 19 E.420-510 N.4634-410.

Orientatio

The orier tion of circular structures is best expressed in terms of the location ard orientation of a planar feature such as a doorway. The original entrane to the Beacon is presently loc its and the plane of te doorway is in a line 6 de or The door of th South Twin is on its n degrees east o magnetic north. Th southwest side and the plane of th of magnetic noth. Orientation

The original orientations of the Three Sisters have been determined within 10°+/- by a careful analysis of Figure 3 and Map 2. The first step in this process was to add to Map 2 the new locations of the Three Sisters thirty feet west of those shown on the 1885 map. Then the points where foreground objects obscure background objects in Figure 3 were plotted on Map 2 and projected to determine the point of view of Figure 3. The locations of storm porches could be determined directly from the photographs and plotted on Map 2 to determine their orientations. The storm porches of the North Twin and the Beacon were originally oriented to the southwest. The line of the doorway of the North Twin was 76° west of magnetic north. The line of the doorway of the Beacon was 63° west of magnetic north. The storm porch of the South Twin is oriented to the northwest and is not visible in Figure 2 but the orientation of the window could be determined and the relation of window to door is shown on sheet 8 for the HSR drawings. The line of the doorway of the South Twin was historically 27° east of magnetic north.

General Description

The three towers are circular wood frame structures covered with wood shingles. The walls taper from a diameter of 14 feet-5-1/8 inches at the base to a diameter (calculated) of 11 feet-6½ inches below the cornice. The taper is approximately 1 inch per foot of height. The diameter of the deck (greatest projection of the cornice) is 14 feet-1½ inches. The shingles are about 16 inches long, exposed to weather 5 inches to 5-3/8 inches and vary from 3 inches to 9½ inches wide. The flat decks on the North and South Twins are covered with mineral-surfaced roll roofing. The deck of the Beacon supports a lantern with a circular parapet and nine plate glass windows. Around the perimeter of the deck is a railing 3 feet-0 inches high supported by eight stanchions (see Sheets 2-5).* The deck and parapet are covered with painted canvas later coated with

---

*Sheet number references are to measured drawings included as Appendix A to this report.

tar. The roof of the lantern is copper sheet with radial standing seams corresponding to the nine glazing bars below. The peak of the roof is capped by a copper ventilator ball with a broken lightning rod above (see Sheet 4).

The three towers have three windows each--two at the first floor and one at the second. Only one window at the first floor of each tower is original (see Figure 10). The evidence for this conclusion is drawn from Figures 3 and 4, paint layering (see Appendix D), and moulding profiles (see Sheet 6). The original windows in the Beacon and the North Twin are to the left of the door. In the South Twin, the original window is to the right of the door. An examination of the historic photographs for window/door relationships shows that the North and South Twins were originally in the north and south positions, respectively, of the Three Sisters and the Beacon was the central tower. The original windows are six-over-six single-hung sash with sash openings of 2 feet-7 inches wide and 4 feet-4¼ inches high. The nonhistoric windows in the Beacon are two-over-two single-hung sash with 2 feet-3-3/4 inches wide and 4 feet-5½ inches high sash openings. The nonhistoric windows in the North and South Twins are six-over-six single-hung sash with 2 feet-6½ inches wide and 4 feet-4 inches high sash openings. There is no firm documentary evidence to date the non-historic windows. The fact that the later windows on the Twins match each other but differ from those on the Beacon strongly suggests that they were installed by the private owners. The most logical time for the private owners to have undertaken this work is in the course of converting the towers to residential use. Therefore a date of c. 1920-1925 is suggested for the non-historic windows. The sash openings are contemporary with the sash.

All three towers retain their original entrance doors (see Figure 11). The historic door for the Beacon was installed in the west wing which was demolished in 1983, but the door was removed and stored on the second floor. The four-panel doors are 2 feet-8 inches by 6 feet-7 and 1-7/8 inches thick. Panel mouldings are the same on all three doors (see Sheet 6). All three towers also retain the original door architraves and

Figure 10

Historic window at the Beacon, typical for all three structures entablatures (see Sheet 9). Another large doorway was cut in the south wall of the Beacon to prôvide access to and from the south wing of the cottage (see Sheet 2).

Documentary evidence presented in this report indicates that the vestibules or "storm porches" which appear on the historic photographs (see Figure 3 and 4) were added to the lighthouses in 1895. The date of their removal is unknown, but it was probably done when the Twins were taken out of service and the Beacon was attached to the keeper's house. The only surviving physical evidence of these storm porches are two vertical lines of sheet metal flashing interleaved with shingles and a gable roof outline surrounding the original door to the Beacon. This evidence combined with evidence from historic photographs and from construction drawings of and existing conditions at the Brant Point Lighthouse (Nantucket Harbor) is sufficient to support a reconstruction of the storm porches, if desired. The Brant Point Lighthouse was built c. 1900 and is very similar to the Three Sisters. (see Figure 12)

Directly under the cornice of the Beacon are eight wooden brackets (see Sheet 3). They were installed in 1911 to hold a steel compression ring to which were attached four galvanized guy wires. The guy wires were required to steady the tower so that winds would not disturb the recently-installed flashing light mechanism (see History Component). Since the Twins never contained flashing light mechanisms, brackets and guy wires were never attached.

Interior Description

Crawl Spaces

The crawl spaces and foundations of the Three Sisters are discussed on p. 88 of this report.

Figure 11

Original entrance door at the Beacon

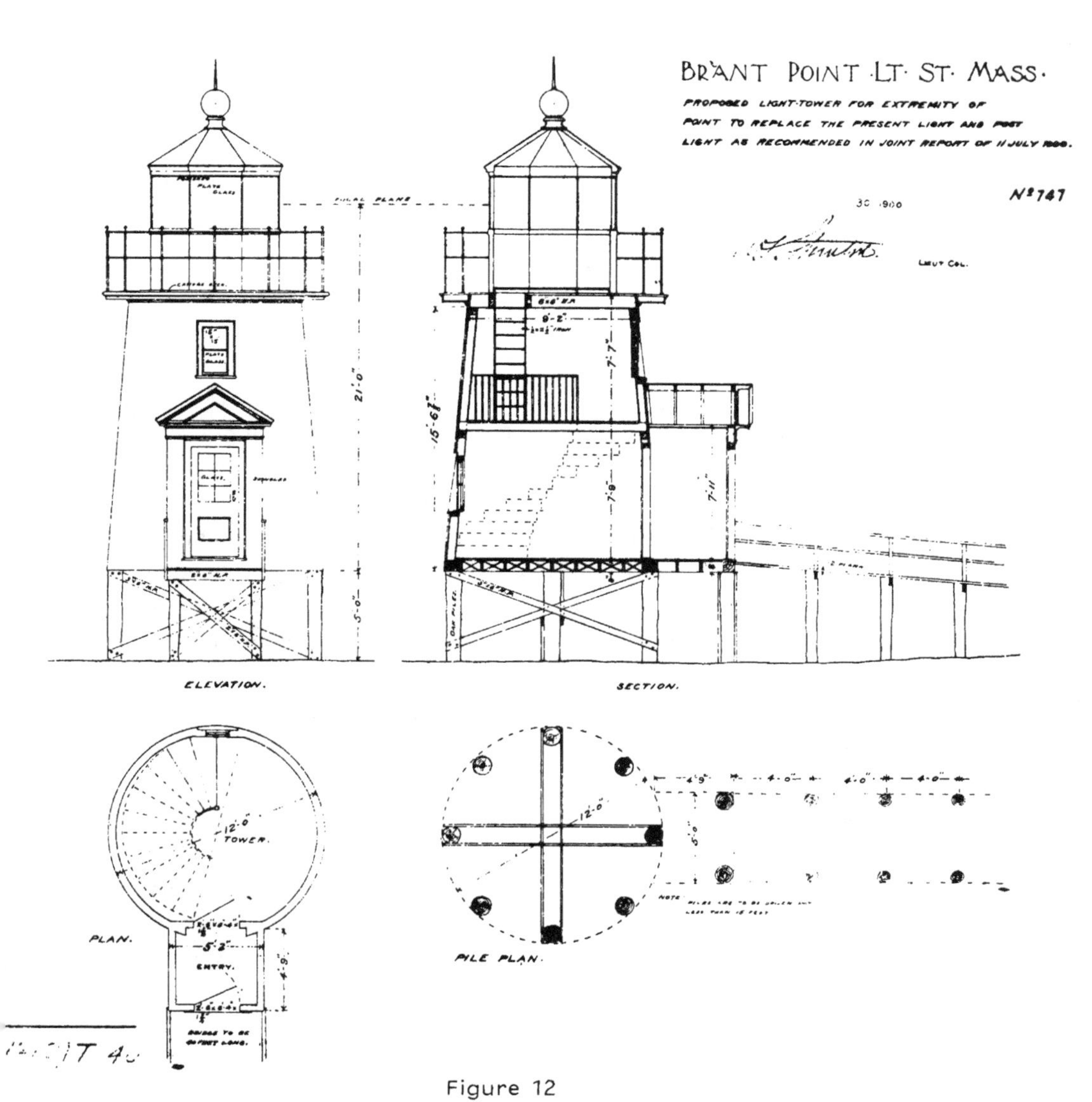

Figure 12

Brant Point Light Station

This lighthouse, erected in 1900-1901 in Nantucket harbor has the most comparable storm porch that could be found.

First Floor

The first floor rooms of all three structures are circular in plan with diameters between 13 feet-0 inches and 13 feet-2½ inches (see Sheets 2, 7, and 8).

Floors

Floors are 3½ inch tongue and groove boards, 7/8 inch thick. There has been one patch in the floor of the Beacon. It is a rectangle 1 foot-6 inches east-west by 1 foot-1 inch north-south located 4 feet-2 inches from the east wall and 5 feet-6½ inches from the north wall. There is a circular hole, 10 inches in diameter, in the subflooring beneath this patch and a joist has been cut and headers installed to support the edges of the opening. The location of the patch in relation to the hole in the lantern floor and the fact that the patch is only present in the Beacon indicate that the hole was probably made to accommodate the fall of the weight which drove the clockwork light-flashing mechanism. There is an 8 inch high baseboard with rectangular profile and no shoe moulding.

Walls

The walls are a sandy-textured two-coat plaster system on circular-sawed wood lath. They rise 10 feet-0 inches to the ceiling and taper inward about 1 inch per foot. The ceiling diameters are approximately 11 feet-7 inches.

Ceilings

The ceilings of all three structures were altered to allow a larger floor area above when they were converted to cottages (see Sheets 2, 6, and 7). The ceiling under the additional floor area in the Beacon is exposed joists and floorboards, unpainted. The original ceiling area of the Beacon retains the same plaster as the walls. Both areas of the ceiling of the North Twin are plastered, but there is a distinct crack between the old and the new and paint is peeling from the new plaster.

The ceiling of the South Twin is fiberboard panels nailed to the joists with wood battens over the joints.

Stairs

All three structures retain the original open-riser stairs (see Sheets 2, 6, 7, and 8; and Figure 13). The stairs in the North and South Twins were moved when the second floors were enlarged. They are light, airy structures, hardly more than ladders, but they make an important contribution to the perception of these rooms as circular spaces. The structural conditions of these stairs are discussed on p. 94 of this report.

Furnishings

A built-in wooden cabinet in the first floor of the Beacon is the only known surviving original furnishing in the Three Sisters (see Figure 14). It is shown on Sheet 6 of the drawings. The door, frame, and drawer fronts seem to be cherry wood. The sides and interior are pine. Presumably it was used for storage of wicks, chimneys, cleaning equipment, tools and supplies which were needed for daily maintenance of the lighting apparatus. There is no evidence of similar cabinets in the other towers. There is extensive evidence of hardware for window coverings on all windows--both historic and nonhistoric. This project did not allow for an exhaustive study of that evidence to determine what, if any, of the evidence relates to historic window coverings. If an interior restoration is to be undertaken, an exhaustive study of that evidence should be included in the construction documents phase. A single drop cord, pullchain light fixture hangs from the center of the ceiling. A duplex outlet is located on the west wall about 18 inches above the floor and adjacent to the north jamb of the original entrance door.

Second Floor

The second floors of the three towers are circles of 11 feet-4 inches to 11 feet-6 inches diameter.

Floors

Flooring in the Beacon is 3½ inch tongue and groove boards, 7/8 inch thick. Flooring in the Twins is 7/8 inch tongue and groove boards of random widths from 5¼ inches to 11½ inches. The original floor areas were semicircles as shown on the floor plans. Stair openings are irregular shapes approximately 7 feet by 3 feet (see Sheets 2, 7, and 8 for floor plans). Railings around the stair openings take different forms. The railing in the Beacon is 5 feet-3¼ inches long and 2 feet-7 inches high. It consists of 2 by 4 inch end posts, a 2 by 4 inch top rail with chamfered edges, and a 1 by 2-3/4 inch intermediate rail. The railing in the North Twin consists of two straight sections 3 feet-6 inches and 2 feet-7¼ inches long and 2 feet-8 inches high. The railing in the South Twin consists of two straight sections, both 3 feet-8 inches high. The section that is attached to the wall is 2 feet-6 inches long, the other is 3 feet-5¼ inches long. The handrail is 2 inches high and 2-3/4 inches wide with a rounded top surface. It is supported on solid panels of 3½ inch beaded tongue and groove boards. The boards are attached at the bottom to the face of the stair opening and covered with a fascia. The joint with the handrail is trimmed with 1/2 inch square stock on both sides.

A 1 by 6 inch plain board serves as baseboard in all three towers. There is no shoe moulding, but the end of the baseboard at the top of the stair was cut at an angle to meet the stair stringer. Where the stairs have been relocated, the cut in the baseboard remains as a clear indicator of its original position. In the South Twin, a similar baseboard has been installed around the edge of the added floor area. No additional baseboard was installed in the Beacon or the North Twin.

Walls

The walls of the three towers are a sandy-textured two-coat plaster on wood lath over wood studs. They are circular and taper to a diameter of 10 feet-5 inches at the bottom of the deck beams, 6 feet-9 inches above the floor.

Figure 13

Stairs at the Beacon

Figure 14

Built-in cabinet at the Beacon

Ceilings

The ceilings of all three towers consist of the underside of the lantern deck, 2-3/4 by 7½ inch grooved and splined planks, and the 8 inch wide by 7½ inch deep circular-sawed beams which support them. Near the center of each ceiling is a board-framed hatch opening which historically provided access to the lantern above. (See Sheet 5) The lantern is still in place on the Beacon, but the lanterns from the Twins were retained by the Bureau of Lighthouses when the towers were sold. The hatches are still in place, however. Roofing material was simply applied over the top of the hatches. Paint lines and nail holes on the hatch frames indicate the locations of the stairs or ladders which provided access to the hatches. There are no corresponding indications of the attachment of ladders to the floors of the lighthouses.

The ceiling of the Beacon retains some 12 inch fiberboard acoustical tiles, stapled to 1 by 3 inch nailers which are in turn nailed to the bottom of the deck beams. The beams and the underside of the deck are unpainted. The joint of this ceiling finish with the wall was trimmed with a ¼ by 2 inch wood strip. The ceiling of the South Twin is the exposed underside of the lantern deck. It has been painted with one coat of white paint. The ragged edge of the plaster is exposed just below the beams. The ceiling of the North Twin is also the exposed underside of the lantern deck, but in this case, unpainted. One of the beams is stenciled "B. James & Co." A cove moulding is applied to the wall below the beams to cover the edge of the plaster. The flat surface of the moulding is not flat against the wall, but the angle between the two flat surfaces is against the wall and the flat surfaces are at an approximate 45 degree angle to the wall. We assume the moulding was applied this way because it would conform more easily to the curvature of the wall.

The two middle lantern deck beams in the South Twin are severely rotted and require replacement.

In all three towers a pattern of nail holes and cut nails at 16 inches on center indicates that there were finished ceilings other than the ones which are present. This ceiling was probably a wood lath and plaster system identical to the walls and the first floor ceiling, but applied to nailers rather than directly to the structure. Further evidence of this historic finish is found on the outside surfaces of the hatch frames in the form of horizontal paint lines where a plaster ceiling would have butted the hatch frames.

Furnishings

The second floor of the Beacon is furnished with a single wall-mounted pull-chain light fixture just west of the window. The wire to this fixture is exposed on the wall down to the baseboard and north to a duplex receptacle in the baseboard. In the western portion of the North Twin second floor is a bracket for a small clothes rod. It is painted first pink, then the present white. The pink is on top of the blue paint on the wall. A similar clothes rod is located on the second floor of the South Twin. It is mounted on a plain board which extends from the baseboard up to 5 feet-6 inches above the floor. There is one duplex outlet in the newer section of baseboard near the end of the original section of baseboard.

The Lantern

Originally, all three towers supported small "lantern" structures whose purpose was to shelter the lighting apparatus from the elements and to allow the greatest possible transmission of light out to sea. As discussed earlier, the lanterns from the Twins were retained by the Bureau of Lighthouses as a source of replacement parts. When the Beacon was sold, its lantern was included, but the lighting apparatus was retained by the Bureau of Lighthouses for use in the iron tower which was moved from Chatham to replace the Beacon. When that lighting apparatus was replaced by the Coast Guard, it was acquired by Cape Cod National Recreation Area and is now on display in the Salt Pond Visitor Center.

The lantern on the Beacon has survived to the present with minimal alterations (see Figure 15). The floor of the lantern consists of 2-3/4 by 7½ inch planks, grooved and joined with splines. The joints have been filled with oakum and caulked like the deck of a ship to prevent water penetration. The planks are face nailed to the supporting beams with large cut spikes which have elongated hexagonal heads with raised circle designs. The nails are countersunk and caulked. Areas of gray paint survive on the floor. This floor is continuous with the lantern deck outside, but that deck has been covered and could not be inspected for this report. The inside diameter of the lantern floor is 6 feet-5¼ inches.

Walls

The walls of the lantern consist of a circular section 3 feet-3½ inches high (the parapet) and a nine-sided glazed area above. The interior finish of the parapet is 3½ inch beaded tongue and groove boards applied vertically to a system of curved nailers at floor level and at 2 feet-8 inches above. The nailers and the structure above are supported by 2 by 2½ inch wood studs, equally spaced around the perimeter both inside and out. The exterior wall is also made of 3½ inch tongue and groove boards. Wall thickness is 1 foot-2 inches. The tops of the boards are flush with the bottom of the cast iron sill at 3 feet-2¼ inches. Twenty-seven of the boards have been removed but are stored in the lantern. Five of these need to be replaced. There are four ventilation intake shafts evenly distributed around the perimeter of the parapet (see sheets 4 and 5). They provided fresh air with minimum draft for the combustion of the illuminating fuel. The ventilator ball on the roof was for exhaust of combustion gases.

The parapet is capped by a three-piece cast iron sill (see Sheet 5). The pieces are identical one-third-of-a-circle arcs with cast-in fittings for joining them together and for attaching the glazing bars to the sill. The pieces of the sill are joined with hexagonal head machine bolts. Glazing bars are attached at the joints and at the third points of the sill pieces for a total of nine glazing bars and nine lights. The sill is 1 foot-2 inches wide and 6 feet-5¼ inches inside diameter with glazing bars and

Figure 15

The lantern at the Beacon

stops mounted near its inner edge. The outer section of the sill is sloped for drainage. The profile of the glazing bars is similar to an airfoil. The outer glass stop is a half-round bar attached with brass machine screws. The remainder of the glazing bar consists of a rabbet to receive the plate glass and two curved surfaces which meet in a vertical line at the inside edge. The connection of the glazing bar to the sill is accomplished with a machine-threaded rod at the bottom of the bar passing through a hole in a lug in the sill and held tight with a hexagonal nut. The joint at the top of the glazing bar is concealed by the eave plate assembly. The lights are 5/16 inch plate glass sheets 2 feet-4 inches wide and 3 feet-0 inches high, but glass stops reduce the clear area to 2 feet-3 inches by 2 feet-11 inches. All of the lights are cracked or broken to varying degrees, probably as the result of inadequate resistance to lateral and torsional forces (see pp. 86-92 of this report for a discussion of the structural performance of the lighthouses).

Ceiling

The ceiling of the lantern is a nine-sided pyramidal shape with a 1'-0 and 3/4" diameter opening at its center. The opening is designed to conduct combustion gases into a chimney and thence into the ventilator ball and to the outside. See Sheet 5 of the drawings. Nine iron bars, 1¼ inches by 1/8 inch, support triangular sheet iron ceiling panels. The manner of attaching the copper roofing to the lantern structure is completely concealed.

Access

Access to the lantern from the tower and from the lantern to the deck or "gallery" is by small hatchways. See Sheet 5 for details. There are two small doors flush with the inner and outer surfaces of the parapet that give access to the deck. Both are curved board and batten doors with the battens on the normally concealed side. The exposed sides are finished to match surrounding materials. The floor hatch is flat and rectangular except for a segment of a circle cutout around the pedestal of the lighting apparatus. The battens are on the bottom of the door.

Furnishings

Roller shade mounting brackets are attached to the eave assembly above each glazing bar. The brackets are offset so that alternate shades will be mounted high and low and the shades will overlap to exclude as much light as possible. Shading was necessary because daylight adversely affects the light transmission characteristics of the lead crystal lens elements. Because of the odd number of lights, two adjacent shades must be mounted at the same height.

Structural Description

The structural elements of the lighthouses will be described in this section. An engineering report on safety and load-bearing limits is presented at the end of this Section. Since it is the most complete of the structures, the Beacon will be the focus of this section.

Lantern

The lantern roof structure is apparently supported on nine radial ribs, 1¼ inches wide, 1/8 inch deep, and 4 feet-3 inches long. The upper sides of the ribs are concealed in the 2 inch thickness of the roof structure and the ribs may actually be the bottom flanges of C-, L-, I-, or Z-shaped beams or even solid bars. The ribs are attached to a compression ring at the center and are supported by the eave assembly at the ends. The eave assembly is concealed, so its exact configuration could not be determined. The visible interior section appears to be a L-shaped iron section. The eave assembly rests on nine lenticular glazing bars with maximum dimensions of 4¼ inches radially and 1½ inches tangentially. The bars are 3 feet-0 inches high and are bolted to a circular cast iron sill that consists of three 120 degree sections with a 6 foot-5¼ inch inside diameter and a 1 foot-2 inch width. This sill is supported by inner and outer circles of 2 by 2½ inch wood studs, six in each circle. The studs rest on curved plates on the lantern deck.

Lantern Deck

The lantern deck consists of 2-3/4 by 7½ inch planks grooved and joined with splines. The diameter of the deck is 14 feet-1½ inches. The planks are supported by four 8 inch wide and 7½ inch deep pine beams at 3 feet-4 inches on center. They in turn are supported on 2 by 4 inch plates atop the circular tower walls.

Walls

The walls of the lighthouse towers consist of 2 by 4 inch studs at 11½ inches on center around the perimeter of the structures and tapering inward approximately 1 inch per foot of height. Sheathing is 7/8 by 2½ inch boards installed at approximately 70 degrees from horizontal. The interior covering is plaster on wood lath.

Second Floor

The composition of the second floor structures was impossible to determine without destructive investigations. Because of the 7½ inch floor thickness and the width and spacing of the first floor joists, it is assumed that the second floor joists are 2 by 6 inches and 16 inches on center. The structure of the additional floor area for the second floor of the Beacon is exposed and consists of 2 by 6 inch joists at 16 inches on center. The maximum span for second floor joists is about 11 feet-8 inches.

First Floor

The first floor structure consists of 2 by 8 inch joists at 16 to 17 inches on center. The maximum span for first floor joists is about 12 feet-8 inches. One of the longer joists near the center of the first floor of the Beacon has been cut and headers have been installed to accommodate the weight for the clockwork flashing mechanism passing through the first floor. The ends of the joists are notched and mortised into an 8 by 8 inch circular wood sill. The sill is built-up from 2 inch dimension lumber, laid flat, trimmed to conform to the circular plan of the lighthouses, and bolted together.

Foundation

The original foundations of the Three Sisters were posts driven 4 feet into the ground.[1] The present foundation of the Beacon is a mix of posts driven an unknown depth into the ground and concrete piers. The sill is anchored to the posts and piers by iron rods. In point of fact, the Beacon was moved on September 13, 1983, and now rests on cribbing near the Twins. This is a temporary storage arrangement, but the tower should, nevertheless, be restrained from lateral movement or overturning by cables and soil anchors as detailed in a DSC memorandum dated February 3, 1984. (See Appendix F)

The foundations of the Twins are circular poured-in-place concrete walls of unknown depth, but probably with a spread footing. It was not possible to investigate the anchoring, if any, of the towers to the foundations because the crawl spaces are inaccessible.

Stairs

The stairs in the Three Sisters consist of curving composite stringers and wedge-shaped treads as shown on Sheet 6 of the drawings. The outer stringers are 1 by 6 inch boards with half-round trim attached to the top surfaces. The inner stringers are also 1 by 6 inch boards but with 1 by 2 inch strips applied to the exposed surface near the top and bottom. This creates a C-shaped cross section that can be analyzed as a beam with a thin web and thicker flanges. The outer stringers are adjacent to the curved exterior walls and may be attached to them inconspicuously if additional support is necessary. The inner stringers span a maximum horizontal distance of 10 feet from first floor to second.

---

1. "Description of Light-House Towers, Buildings, and Premises at Nauset Beach, Massachusetts, Light-Station, January 11, 1910," pp. 2-3, First District Coast Guard Files, Boston, Massachusetts.

The treads are of constant 1-1/8 inch thickness and widths vary from 6¼ inches at the inner stringers to 10 inches at the outer stringers.

Load-Bearing Limits of Structure

The existing conditions and proposed uses of the three lighthouses are contained in the architectural data section. As stated, the Beacon is the only lighthouse of the three designated for public visitation. While the new foundation will be similar for all three lighthouses, the load analysis of existing walls and floors will be based on those conditions existing at the Beacon. The existing structural systems of the Twins appear stable for their proposed use.

Foundation

The proposed foundation at the new location of the Three Sisters is a continuous concrete wall and spread footing. The foundation will be designed for a net bearing capacity of 4,000 psf as recommended in the "Geotechnical Engineering Report" by Eastern Geotechnical Associates, December 21, 1982. Footing dimensions will be determined by imposed dead loads and floor live loads as noted in the floor section below. The lighthouses will be anchored to the foundation to prevent overturning due to wind loading. The proposed foundation is shown on the drawings.

Floor Framing

The floor capacities shown below are derived from existing conditions at the Beacon. Because many structural members are hidden from view with architectural finishes, the analysis is based on sound members with full cross-section. Two lantern deck beams in the South Twin are severely rotted and should be replaced. If, during further investigation, other structural members are found to be decayed, then additional analysis and/or repair is required.

The Beacon Floor Load Analysis

Floor Level	Floor Framing	Allowable Total Load (psf)	Existing Dead Load (psf)	Allowable Live Load (psf)
First floor	2x8 c 16" o.c.	53	11	42
Second floor	2x6 c 16" o.c.	31	7	24
Lantern deck	8x8 c 40" o.c.	149	69*	80

*Lantern bears on deck beams. Dead load of lantern estimated to be 1500 pounds. Includes 25 psf snow load on lantern roof.

For comparison purposes the minimum uniformly distributed live load, as listed in the 1980 Massachusetts State Building Code, for places of assembly (public use) is 100 psf. Using this criteria, the designated public assembly areas (first and second floors) will not meet minimum live load requirements. Due to potential for floor overload, the floor systems must be either strengthened to accommodate the additional loading or use of the space must be limited. In the latter case, this requires limiting the number of people in the structure at one time. This would appear to be the preferred alternative because strengthening the existing floor systems would damage the architectural finishes. Also, the size of the building alone provides limitation in terms of number of occupants. Based on the existing framing, the maximum number of people permissable at the first floor is 10 persons and at the second floor is 5 persons. Loading beyond this number will require strengthening of the floor systems.

Lantern Roof

The metal roof framing at the Beacon is inaccessible, therefore structural member sizes could not be determined. In order to determine the load bearing capacity, the member sizes and condition must be known. It can be noted, however, that the lantern roof has existed for 130 years with no visible deterioration. Investigation to determine the existing member sizes and condition at the Beacon will require some destruction of historic fabric. This investigation should be performed if the lantern is opened for public use.

If reconstructed lantern roofs are installed on the Twins, then the roof framing should be designed for contemporary loading. In accordance with the current Massachusetts State Building Code, the design roof live load (snow) is 25 psf. No difficulties are anticipated in the roof framing design for this value. No attempt will be made to reproduce the roof framing of the Beacon, because the framing will be concealed and investigation to determine the lantern roof framing over the Beacon would destroy historic fabric.

Wall Framing

This part is concerned with gravity load imposed on the walls only. (The lateral analysis of the walls is contained in the following part.) The stud wall framing consists of 2 x 4"s at 11½ o.c. with wood sheathing. The wood sheathing acts as continuous lateral support for each stud. Based on this framing situation, the existing wall framing can support a 100 psf live load on both the second floor and lantern deck.

Lateral Analysis

The existing stud wall framing is capable of resisting the 21 psf wind load required by the Massachusetts State Building Code (0-50 feet building height, wind load zone 3, Exposure C). This load controls for light-weight, low-height buildings instead of earthquake load in this area. The wall sheathing acts as a diaphragm to transfer the wind load to the foundation. Anchoring the lighthouses at the foundation against overturning has been mentioned previously in the Foundation part.

Some cracking has appeared in the window area of the Beacon lantern. This appears to be caused by movement of the metal window framing and the lack of lateral rigidity. Some stiffening of the existing members may be required to protect breakage of new glass. If new lanterns are reconstructed for the Twins, the members will be designed with adequate stiffness against wind load.

Summary

The existing lighthouses are in basically good structural condition Two problems occur at the Beacon which are potential overload of the floor levels and lack of information needed to properly analyse the lantern roof framing. Overload of the floors can be resolved by limiting the number of people per floor level. Additional investigation of the lantern roof framing should be performed if the lantern is opened to the public.

Proposed Work

Treatment Alternatives

It is our firm conviction that the treatments proposed and carried out on the Three Sisters should be consistent with NPS policy as expressed in NPS-28. We also believe that the treatments should be internally consistent, i.e., a preservation alternative should not include any restoration or reconstruction work that is not absolutely essential to the stabilization and preservation of the structure.

The task directive for this project identified five alternative approaches to the treatment of the Three Sisters. Those alternatives were formulated as treatments for the missing elements of the lighthouses and they are listed directly below. For this report, the alternatives have been reformulated as treatments for the three structures in their existing condition. The HSR Alternatives are listed after the Task Directive Alternatives with corresponding numbers. The use of substitute materials is now an option to be considered in the construction documents phase of the Restoration alternatives.

TASK DIRECTIVE ALTERNATIVES

1. No action

2. Historically accurate reconstruction

3. Approximate reconstruction using modern materials and techniques

4. Installing similar (or identical) elements salvaged from other sites

5. "Ghosting" the missing element with a framework

HSR alternatives

1. Preservation

2. Restoration

3. deleted

4 Restoration using salvaged elements

5. "Ghosting" the missing elements with a framework

<u>Preservation</u>

It must be noted that a pure preservation and stabilization treatment was made impossible by the demolition of the attached cottage structures and the move of the Beacon. The following items of work are necessary to preserve the Three Sisters:

1. Repair or replace roofing on the Twins. Replace the deck covering on the Beacon.

2. Repair or replace cornice mouldings as necessary.

3. Repair all existing windows. (This work would be extensive for some of the later windows.)

4. Completely reshingle all three towers.

5. Repair sheathing as necessary at the time of reshingling.

6. Repair entrance door frames. Install flashing at top of door frames.

7. Repair termite-damaged first floor joists and sills.

8. Replace two rotted deck beams in the south Twin.

9. Install new framing, sheathing, and shingles in the nonhistoric door opening on the south side of the Beacon. (This work can be justified as stabilization on the basis of the lack of an existing closure for this opening and the necessity for such a closure to prevent weather damage to the structure. The opening was exposed as a result of the removal of the cottage wings.)

10. Repair wood parapet inside and outside.

11. Replace all plate glass in lantern.

12. Clean and repaint all metal surfaces of the lantern.

13. Anchor the Beacon against lateral and uplift forces. (This may be accomplished in a temporary manner with cables and soil anchors or permanently with a new foundation.)

14. Repaint all towers in present color schemes.

The following items of work would be inappropriate for the preservation alternative because they constitute restoration or reconstruction treatments:

1. Remove nonhistoric windows and install new framing, sheathing and shingles over the openings. This work would be less costly on both an initial cost and a maintenance cost basis than repairing the existing windows, but it is philosophically inappropriate. If preservation treatment is undertaken as an interim measure prior to restoration, then it would be appropriate to remove the nonhistoric windows and shingle over them.

2. Move the towers to the proposed new site and locate them in a line at 150 feet apart. Though this would accomplish the objective of making the lighthouses available for visitor enjoyment and interpretation, it would be inconsistent to restore the towers to a circa 1910 (or earlier) site arrangement without restoring the towers to their appearance at that time. (See sheet 11, Alternative B of the HSR drawings for a graphic representation of this proposal.)

3. Reconstruct storm porches for all three towers and lanterns for the Twins. This work is clearly incompatible with a preservation treatment and would complicate a future move to the proposed new site.

Restoration

The project team recommends restoration of the lighthouses. There is sufficient documentary, physical, and photographic evidence for a restoration to be done with a minimum of conjecture. Since a restoration treatment by definition involves "recovering the historic form and details of a structure," we must first determine which of the many appearances over time of the Nauset Beach Light Station is the "historic" appearance.

We will approach the problem of identifying the historic period for the Three Sisters by progressively narrowing the possible range of dates for a restoration. The earliest possible date is 1892, the year of construction. The latest possible date is 1923 when the last of the Three Sisters ceased being used as a lighthouse. Since the significance of the Three Sisters is due in large part to their number, site plan and use as lighthouses, we can pinpoint the end of the historic period at June 1, 1911, when the Beacon became a flashing light (attached to the keeper's residence) and the Twins were extinguished. The only change to the lighthouses between 1892 and 1911 was the addition of storm porches in 1895. The longer time span, better documentation, and preservation philosophy all support the selection of 1895 to 1911 as the historic period. The period 1895 to 1911 includes sixteen of nineteen possible years; we have the official report of the inspection of the lighthouses for 1910; and preservation philosophy would prefer saving the evidence of the storm porches in place in a restoration treatment to destroying that evidence to restore the lighthouses to an earlier period. This treatment is illustrated as Alternative A on Sheet 11 of the HSR drawings.

The following items of work are necessary for and consistent with a restoration treatment:

1. Move all three towers to the proposed new site and arrange them in a line at 150 feet on center on new foundations.

2. Reconstruct lanterns for the Twins using the Beacon lantern as a model.

3. Reconstruct storm porches at the entrances to all three towers using physical evidence, historic photographs, and comparative evidence of the Brant Point Lighthouse

4. Replace roofing on the lantern decks and parapets of all three lighthouses.

5. Repair or replace cornice mouldings as necessary.

6. Completely reshingle all three towers.

7. Repair sheathing as necessary at the time of reshingling.

8. Repair entrance door frames.

9. Repair termite-damaged first floor joists and sills.

10. Replace two rotted deck beams in the south Twin.

11. Install new framing, sheathing, and shingles in the nonhistoric door opening on the south side of the Beacon and in the six nonhistoric window openings.

12. Repair the historic windows.

13. Repair wood parapet inside and outside.

14. Replace all plate glass in lantern.

15. Clean and repaint all metal surfaces of the lantern on the Beacon.

16. Replace the stanchions and railings on all three lighthouses.

17. Repaint all three lighthouses in their historic color schemes.

18. Remove brackets from the Beacon.

If the restoration alternative is selected for implementation, but funds are not available to accomplish the entire job at one time, the work could be performed in two phases. The first phase would be a preservation effort including items 6, 7, 10, 11, 13, and 14 above. A later restoration phase should include items 1, 2, 3, 4, 5, 8, 9, 12, and 15 above. Modern materials may be considered as substitutes for cast iron on the final designs for reconstruction of lanterns for the Twins.

Restoration Using Salvaged Elements

A review of NPS-28 shows the use of similar salvaged elements to be improper.

> The reproduction of missing elements made with new, similar, or substitute materials must duplicate the composition, design, color, texture, and other visual qualities of the missing architectural features. Their design is to be substantiated by historical, physical, or pictorial evidence rather than based on

conjectural designs or the availability of different architectural features from other structures.[2]

It would also be improper to use identical elements unless salvage was the only possible way to preserve them. The only known identical lantern is on the Aransas Pass Lighthouse and it is well maintained. No identical storm porches are known.

Therefore this alternative is inappropriate.

"Ghosting" the Missing Element with a Framework

"Ghosting" as a treatment alternative is not addressed in NPS-28. It has been done successfully on two structures managed by the National Park Service--the Benjamin Franklin House and the gunboat "Cairo." Both of these projects have been recognized with several design awards--most recently a Presidential Design Award for Franklin Court. We believe "ghosting" is an appropriate solution to the dilemma of restoring a structure to meet interpretive goals when there is insufficient information for a restoration with minimal conjecture. It may also be an appropriate solution to the dilemma of how to treat a structure for which there is sufficient information to justify a restoration, but insufficient funding to accomplish it. We consider the information presented in this report sufficient to support a restoration treatment, but we have also provided a cost estimate for "ghosting" lanterns and storm porches so that management may consider this alternative.

Life Safety and Property Protection

Given the small size of the Beacon, the limited sources of ignition present, and the limits on occupancy imposed by its structural capacity,

---

2. Cultural Resources Management Guidelines, NPS-28, Release No. 2, December 1981, Chapter 2, Page 7.

minimal measures should suffice to protect the lives of its occupants. The limit of five persons on the second floor and ten persons on the first should be strictly enforced for life safety as well as structural reasons. The single exit is the controlling factor for first floor occupancy, and the availability of the lantern deck and parapet hatches for emergency escape provide a sufficient level of safty to allow occupancy of the second floor. Ionization detectors, one for each floor with local signal, are recommended to enhance life safety. Because the Twins will not be occupied, no life safety measures will be necessary.

For property protection, fire suppression equipment should be available in each tower. If water supply to the site can be provided, a fire hydrant (or two) and a fire cache/hose house with sufficient hose and other firefighting equipment is recommended. The fire cache/hose house could be located at the parking lot (requiring a great length of hose to reach the north tower) or at the edge of the clearing near the central tower or two hose houses could be located midway between the central tower and the north and south towers.

Accessibility

Because the Beacon is the only one of the three structures which may be open for visitor use, it is the only one which must be made accessible to visitors with impaired mobility. The floor of the sotrm porch will be level with the first floor. The historic photographs do not show whether stairs or ramps led to the storm porches; therefore, the Beacon may be set close enough to grade that the path can slope directly to the threshold. Thus, the first floor can be made accessible, but the steep stair and limited floor area make access to the second floor impossible for those with impaired mobility. An alternative interpretive program should be devised for the mobility-impaired.

Analysis of Effects

Preservation

A preservation treatment would have the minimum impact on the physical fabric of the lighthouses. It would not allow the desired visitor experience of seeing the Three Sisters in a row and as they appeared between 1895 and 1910. It would also require consultation with the State Historic Preservation Officer and repetition of the Section 106 compliance process.

Restoration

A restoration to the historic period, 1895 to 1911, would require removal of nonhistoric elements and reconstruction of missing historic elements. It would provide the optimal visitor experience of seeing the Three Sisters together in their approximate historic appearance, but without the other structures of the Nauset Beach Light Station and on a different site with different compass orientation. The new site also seems to present some difficulties in aligning the focal planes of the three lights at the same elevation with respect to each other. The original elevation of 97 feet above sea level cannot be achieved at the new site. If this treatment is implemented, these shortcomings should be remedied by interpretive means. Since this treatment was included in the DCP which was reviewed and approved by the SHPO and the Advisory Council on Historic Preservation, it can be accomplished within the parameters of the existing PMOA.

"Ghosting"

"Ghosting" of the missing lanterns and storm porches would require removal of some nonhistoric windows and reconstruction of lanterns and porches with an open framework. This alternative would suggest to the visitor what the historic appearance was without recreating it precisely. It would share the shortcomings of the restoration alternative listed above and would require consultation with the State Historic Preservation Officer and repetition of the Section 106 compliance process.

	NAUSET BEACH LIGHTHOUSES HSR CONSTRUCTION COST ESTIMATES FOR PROPOSED TREATMENTS	QUANTITY	UNITS	COST/UNIT	PRESERVATION	RESTORATION	GHOSTING	OPTIONS
1	General requirements	(included in individual work items)						
.	Site work							
	Clearing		lump sum			1500.00	1500.00	
	Foundation excavation	150.00	lin. ft.	5.00		750.00	750.00	
	Water line	500.00	lin. ft.	12.00		6000.00	6000.00	
	Underground electrical line	500.00	lin. ft.	7.00		3500.00	3500.00	
	Walks	222.00	sq. ft.	20.00		4440.00	4440.00	
	Move three towers	360.00	sq. ft.	12.00		4320.00	4320.00	
	Remove nonhistoric windows	6.00	each	50.00		300.00	300.00	
. .	Concrete							
	Footings & stem walls	150.00	lin. ft	60 00		9000 00	9000 00	
.	Masonry	none						
.	Metals							
	Cast iron lanterns	2.00	each	7500.00		15000 00		
	Galvanized steel lanterns	2.00	each	5000.00				
	Stainless steel lanterns	2.00	each	6000.00			5000 00	
	Replace stanchions and rails	400.00	lin. ft.	7.00		1200 00		
.	Wood & plastics							
	Replace deck beams	2 00	each	500.00	1000.00	1000.00	1000.00	
	Replace lantern deck planks	20 00	sq. ft.	3.50	70.00	70.00	70.00	
	Repair first floor joists	100 00	lin. ft.	2.00	200.00	200.00	200.00	
	Infill Beacon doorway	42.00	sq. ft.	15.00	630.00	630.00	630.00	
	Infill nonhistoric windows	48 00	sq. ft.	5.00		240.00	240.00	
	Reconstruct storm porches complete	60.00	sq. ft.	30.00		1800.00	900.00	
	Repair or replace cornice moldings	126.00	lin. ft.	8.00	1008.00	1008.00	1008.00	
	Repair parapet inside & out	30.00	lin. ft.	11.00		330.00	330.00	
	Miscellaneous carpentry repairs		lump sum		500.00	500.00	500.00	
7	Thermal & moisture protection							
	Reshingle three towers	2500.00	sq. ft.	4.00	10000.00	10000.00	10000.00	
	Repair sheathing as necessary	2500.00	sq. ft.	0.85	2125.00	2125.00	2125.00	
	Repair roofing on Twins	260.00	sq. ft.	0.50	130.00			
	Replace roofing on Twins	260.00	sq. ft.	1.00		260.00	260.00	
	Replace roofing on Beacon	130.00	sq. ft.	1.00	130.00	130.00	130.00	
	Repair Beacon lantern roof		lump sum			400.00	400.00	
	New lantern roofs for Twins	90.00	sq. ft.	20.00		1600.00	0.00	
.	Doors & windows							
	Repair nonhistoric windows	6 00	each	500.00	3000.00			
	Repair historic windows	3.00	each	500.00	1500.00	1500.00	1500.00	
	Replace plate glass· Beacon lantern	72 00	sq. ft.	15.00	1080.00	1080.00	1080.00	
	Install plate glass in new lanterns	144.00	sq. ft.	15.00		2160.00		
	Repair entrance door frames	3 00	each	250.00	750.00	750.00	750.00	
	Finishes							
	Prepare & prime exterior wood	2700.00	sq. ft.	0.40	1080.00	1080.00	1080.00	
	Paint shingles	2500.00	sq. ft.	0.40	1000.00	1000.00	1000.00	
	Paint windows & doors	6.00	each	60.00	720.00	760.00	760.00	
	Prepare & prime metals	120.00	lin. ft.	0.75	90.00	90.00	90.00	
	Paint metals	120.00	lin. ft.	0.75	90.00	90.00	90.00	
	Repair plaster	100.00	sq. ft.	3.50		350.00	350.00	
	Paint int walls & ceilings	900.00	sq. ft.	0.30		270.00	270.00	
	Paint int. trim	120.00	lin. ft.	1.60		192.00	192.00	
10.	Specialities	none						
11	Equipment	none						
12.	Furnishings	none						
13.	Special construction	none						
14.	Conveying systems	none						
15	Mechanical							
	Fire hose house (optional)	1.00	each	4000.00				4000.00
	Sprinkler system (optional)	600.00	sq. ft.	6.00				3600 00
16.	Electrical							
	Fire detection system (optional)	600.00	sq. ft.	2.00				1200.00
	Intrusion detection system (optional)	600.00	sq. ft.	2.00				1200.00
	Building wiring	600.00	sq. ft.	5.00		3000.00	3000.00	
	Telephone service		lump sum			1500.00	1500.00	
					25103.00	78225.00	62365.00	10000.00

APPENDIX A
MEASURED DRAWINGS

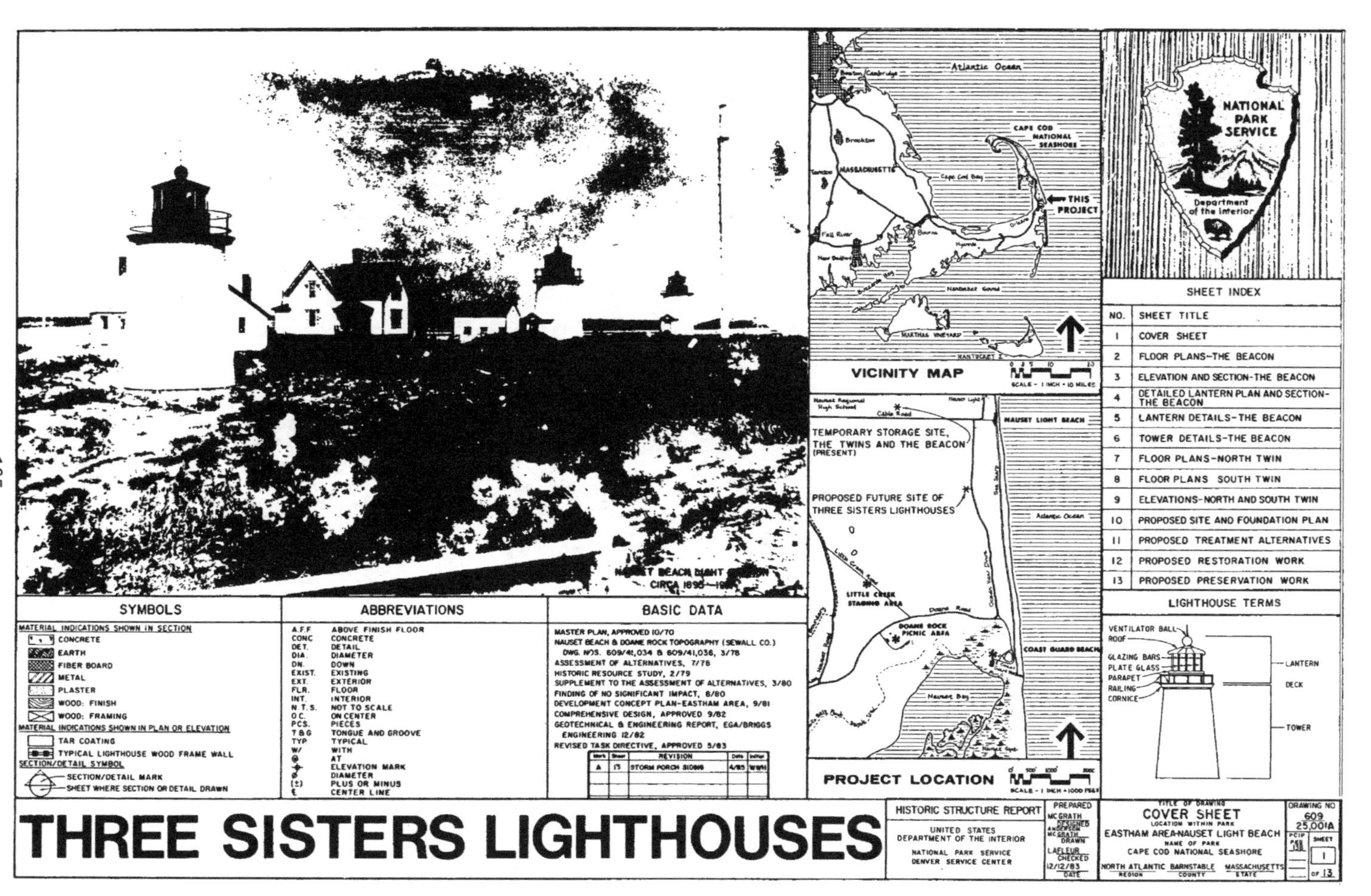

NATIONAL PARK SERVICE
Department of the Interior
Atlantic Ocean
CAPE COD NATIONAL SEASHORE
THIS PROJECT
Cape Cod Bay
MASSACHUSETTS
Brockton
Fall River
New Bedford
Hyannis
Nantucket Sound
MARTHAS VINEYARD
NANTUCKET
VICINITY MAP
SCALE - 1 INCH = 10 MILES
NAUSET LIGHT BEACH
Cable Road
TEMPORARY STORAGE SITE, THE TWINS AND THE BEACON (PRESENT)
PROPOSED FUTURE SITE OF THREE SISTERS LIGHTHOUSES
Atlantic Ocean
LITTLE CREEK STAGING AREA
DOANE ROCK PICNIC AREA
Doane Road
COAST GUARD BEACH
Nauset Bay
PROJECT LOCATION
SCALE - 1 INCH = 1000 FEET
SHEET INDEX
NO. | SHEET TITLE
1 | COVER SHEET
2 | FLOOR PLANS-THE BEACON
3 | ELEVATION AND SECTION-THE BEACON
4 | DETAILED LANTERN PLAN AND SECTION-THE BEACON
5 | LANTERN DETAILS-THE BEACON
6 | TOWER DETAILS-THE BEACON
7 | FLOOR PLANS-NORTH TWIN
8 | FLOOR PLANS SOUTH TWIN
9 | ELEVATIONS-NORTH AND SOUTH TWIN
10 | PROPOSED SITE AND FOUNDATION PLAN
11 | PROPOSED TREATMENT ALTERNATIVES
12 | PROPOSED RESTORATION WORK
13 | PROPOSED PRESERVATION WORK
LIGHTHOUSE TERMS
VENTILATOR BALL
ROOF
GLAZING BARS
PLATE GLASS
PARAPET
RAILING
CORNICE
LANTERN
DECK
TOWER
SYMBOLS
MATERIAL INDICATIONS SHOWN IN SECTION
CONCRETE
EARTH
FIBER BOARD
METAL
PLASTER
WOOD: FINISH
WOOD: FRAMING
MATERIAL INDICATIONS SHOWN IN PLAN OR ELEVATION
TAR COATING
TYPICAL LIGHTHOUSE WOOD FRAME WALL
SECTION/DETAIL SYMBOL
SECTION/DETAIL MARK
SHEET WHERE SECTION OR DETAIL DRAWN
ABBREVIATIONS
A.F.F. ABOVE FINISH FLOOR
CONC CONCRETE
DET. DETAIL
DIA. DIAMETER
DN. DOWN
EXIST. EXISTING
EXT. EXTERIOR
FLR. FLOOR
INT. INTERIOR
N.T.S. NOT TO SCALE
O.C. ON CENTER
PCS. PIECES
T&G TONGUE AND GROOVE
TYP. TYPICAL
W/ WITH
@ AT
ELEVATION MARK
DIAMETER
(±) PLUS OR MINUS
CENTER LINE
BASIC DATA
MASTER PLAN, APPROVED 10/70
NAUSET BEACH & DOANE ROCK TOPOGRAPHY (SEWALL CO.)
DWG. NOS. 609/41,034 & 609/41,036, 3/78
ASSESSMENT OF ALTERNATIVES, 7/78
HISTORIC RESOURCE STUDY, 2/79
SUPPLEMENT TO THE ASSESSMENT OF ALTERNATIVES, 3/80
FINDING OF NO SIGNIFICANT IMPACT, 8/80
DEVELOPMENT CONCEPT PLAN-EASTHAM AREA, 9/81
COMPREHENSIVE DESIGN, APPROVED 9/82
GEOTECHNICAL & ENGINEERING REPORT, EGA/BRIGGS ENGINEERING 12/82
REVISED TASK DIRECTIVE, APPROVED 5/83
REVISION
A | 13 | STORM PORCH SIDING | 4/85
THREE SISTERS LIGHTHOUSES
HISTORIC STRUCTURE REPORT
UNITED STATES DEPARTMENT OF THE INTERIOR
NATIONAL PARK SERVICE
DENVER SERVICE CENTER
PREPARED
MCGRATH
DESIGNED
DRAWN
LAFLEUR
CHECKED
12/12/83
DATE
TITLE OF DRAWING
COVER SHEET
LOCATION WITHIN PARK
EASTHAM AREA-NAUSET LIGHT BEACH
NAME OF PARK
CAPE COD NATIONAL SEASHORE
NORTH ATLANTIC
REGION
BARNSTABLE
COUNTY
MASSACHUSETTS
STATE
DRAWING NO
609
25,001A
SHEET
1
OF 13

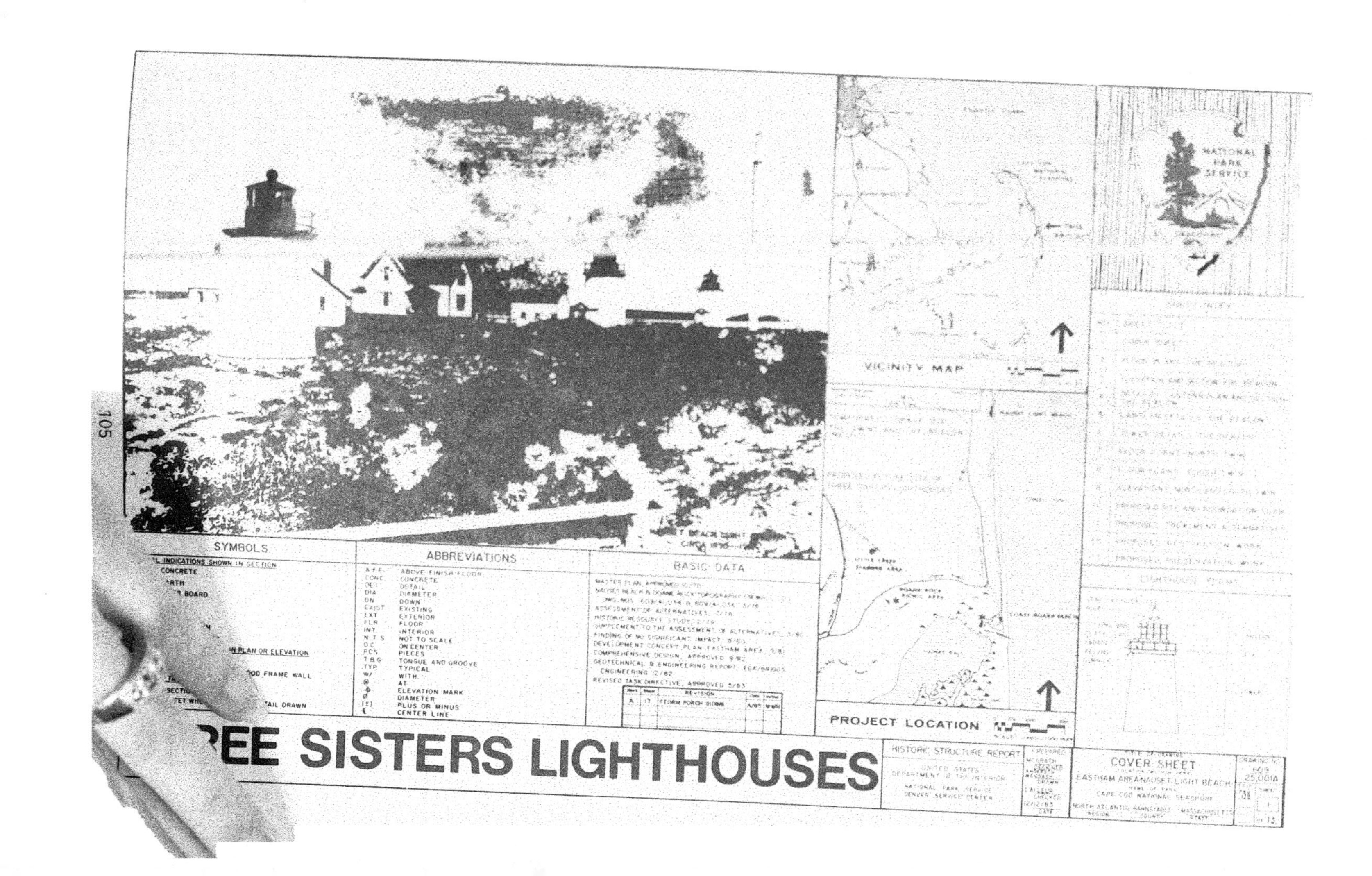

SYMBOLS
ABBREVIATIONS
A.F.F. ABOVE FINISH FLOOR
CONC CONCRETE
DET DETAIL
DIA DIAMETER
DN DOWN
EXIST EXISTING
EXT EXTERIOR
FLR FLOOR
INT INTERIOR
N.T.S. NOT TO SCALE
O.C. ON CENTER
PCS PIECES
T&G TONGUE AND GROOVE
TYP TYPICAL
W/ WITH
@ AT
ELEVATION MARK
DIAMETER
(±) PLUS OR MINUS
CENTER LINE
BASIC DATA
VICINITY MAP
PROJECT LOCATION
NATIONAL PARK SERVICE
EE SISTERS LIGHTHOUSES
HISTORIC STRUCTURE REPORT
UNITED STATES DEPARTMENT OF THE INTERIOR
NATIONAL PARK SERVICE
DENVER SERVICE CENTER
COVER SHEET
EASTHAM AREA NAUSET LIGHT BEACH
CAPE COD NATIONAL SEASHORE
609
25,001A

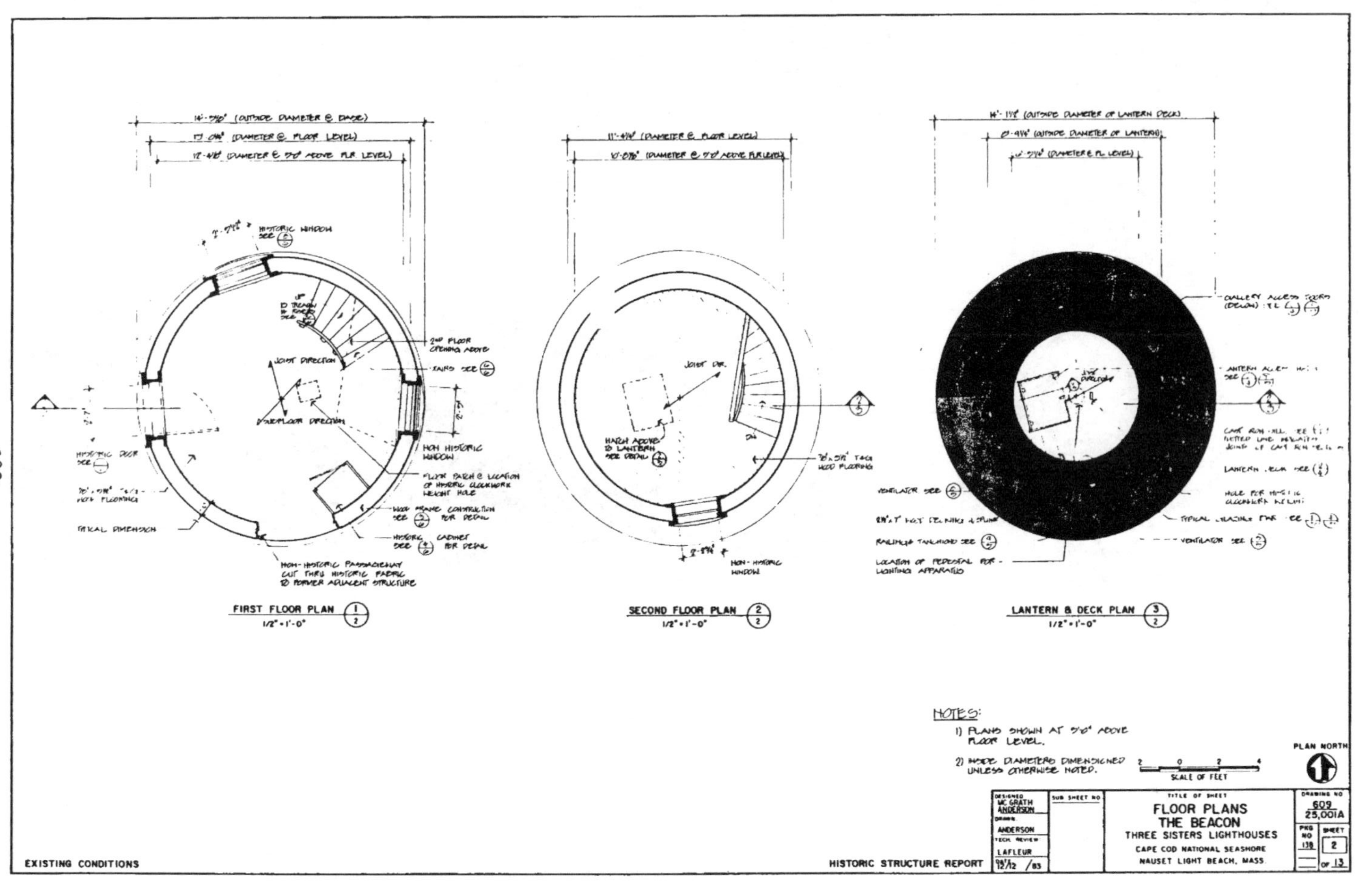
FIRST FLOOR PLAN 1/2
1/2" = 1'-0"
SECOND FLOOR PLAN 2/2
1/2" = 1'-0"
LANTERN & DECK PLAN 3/2
1/2" = 1'-0"
HISTORIC WINDOW
JOIST DIRECTION
2ND FLOOR OPENING ABOVE
NON HISTORIC WINDOW
HISTORIC DOOR
TYPICAL DIMENSION
NON-HISTORIC PASSAGEWAY CUT THRU HISTORIC FABRIC TO FORMER ADJACENT STRUCTURE
JOIST DIR.
HATCH ABOVE TO LANTERN SEE DETAIL
NON-HISTORIC WINDOW
VENTILATOR SEE
LOCATION OF PEDESTAL FOR LIGHTING APPARATUS
GALLERY ACCESS DOOR (BELOW)
HOLE FOR HISTORIC CLOCKWORK WEIGHT
TYPICAL GLAZING BAR
NOTES:
1) PLANS SHOWN AT 5'0" ABOVE FLOOR LEVEL.
2) INSIDE DIAMETERS DIMENSIONED UNLESS OTHERWISE NOTED.
SCALE OF FEET
PLAN NORTH
DESIGNED MC GRATH ANDERSON
DRAWN ANDERSON
TECH. REVIEW LAFLEUR
12/12 /83
SUB SHEET NO
TITLE OF SHEET
FLOOR PLANS
THE BEACON
THREE SISTERS LIGHTHOUSES
CAPE COD NATIONAL SEASHORE
NAUSET LIGHT BEACH, MASS.
DRAWING NO 609 25,001A
PKG NO 138
SHEET 2 OF 13
EXISTING CONDITIONS
HISTORIC STRUCTURE REPORT

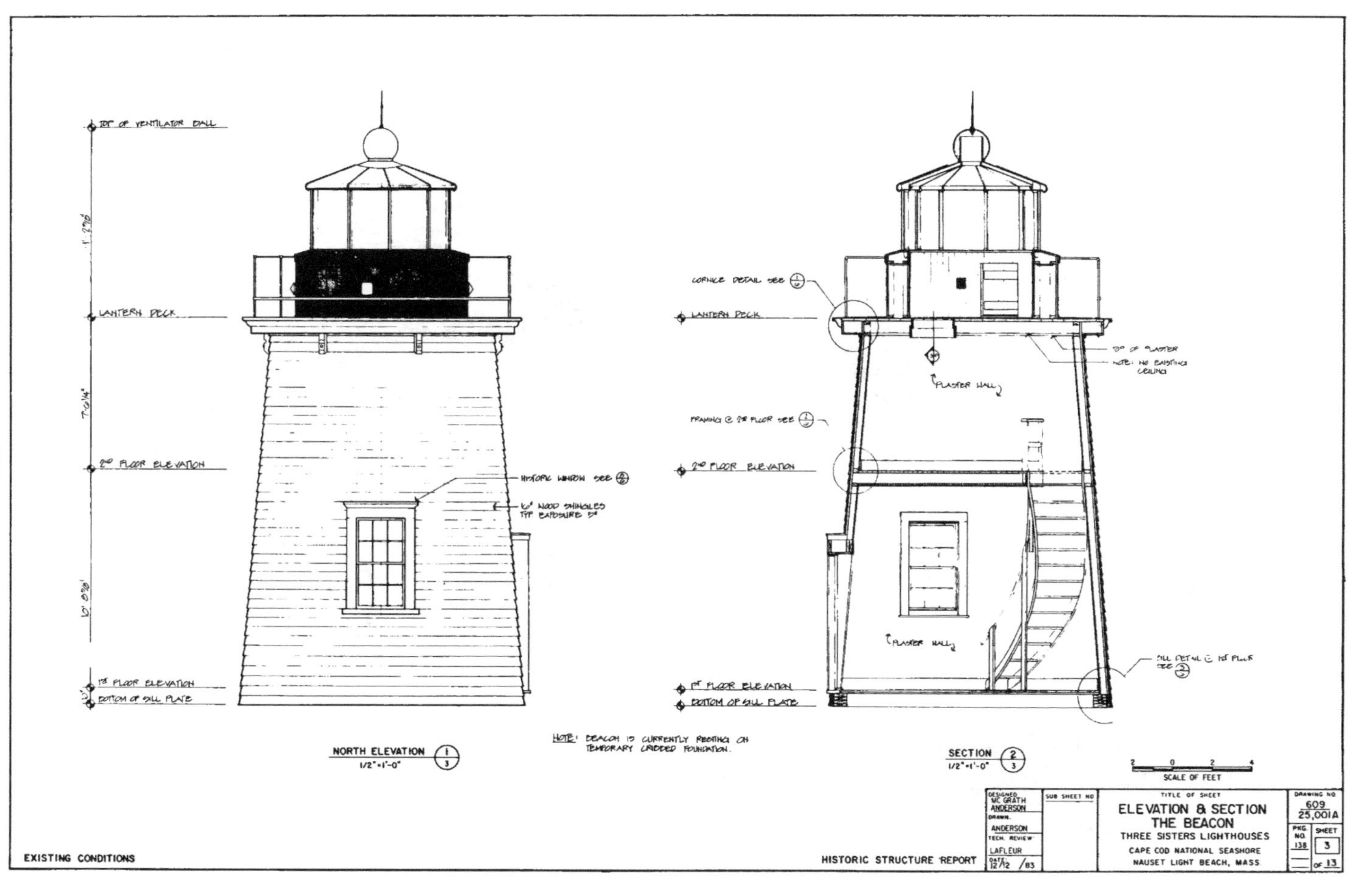

TOP OF VENTILATOR BALL
LANTERN DECK
2ND FLOOR ELEVATION
1ST FLOOR ELEVATION
BOTTOM OF SILL PLATE
HISTORIC WINDOW SEE
6" WOOD SHINGLES TYP EXPOSURE 5"
CORNICE DETAIL SEE
TOP OF PLASTER
NOTE: NO EXISTING CEILING
PLASTER WALL
FRAMING @ 2ND FLOOR SEE
SILL DETAIL @ 1ST FLOOR SEE
NOTE: BEACON IS CURRENTLY RESTING ON TEMPORARY CRIBBED FOUNDATION.
NORTH ELEVATION
1/2" = 1'-0"
SECTION
1/2" = 1'-0"
SCALE OF FEET
DESIGNED MC GRATH ANDERSON
DRAWN ANDERSON
TECH. REVIEW LAFLEUR
DATE 12/12/83
SUB SHEET NO
TITLE OF SHEET
ELEVATION & SECTION
THE BEACON
THREE SISTERS LIGHTHOUSES
CAPE COD NATIONAL SEASHORE
NAUSET LIGHT BEACH, MASS.
DRAWING NO 609 25,001A
PKG. NO. 138
SHEET 3 OF 13
EXISTING CONDITIONS
HISTORIC STRUCTURE REPORT

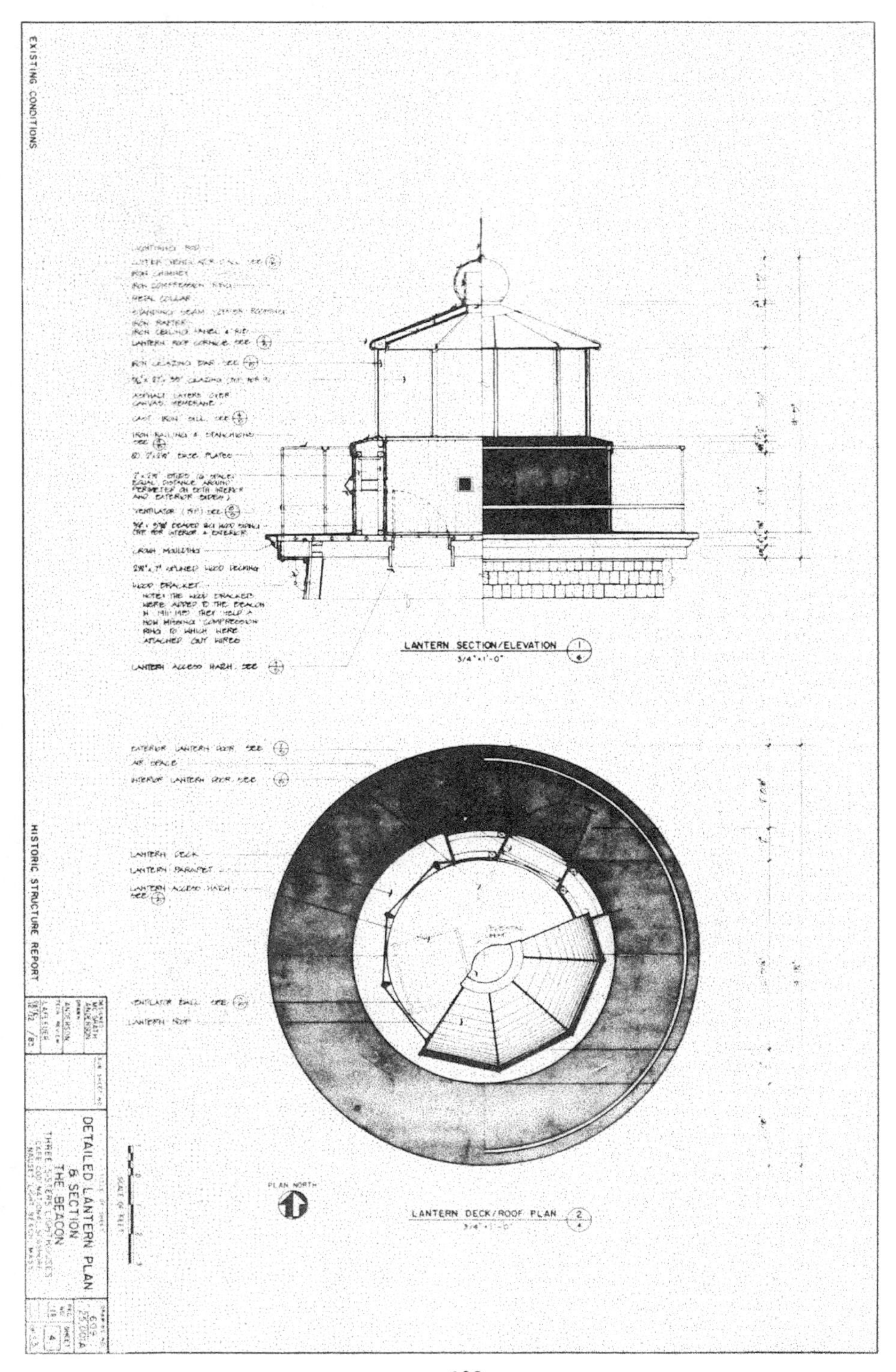
LANTERN SECTION/ELEVATION
3/4"=1'-0"
LANTERN DECK/ROOF PLAN
3/4"=1'-0"
PLAN NORTH
DETAILED LANTERN PLAN
& SECTION
THE BEACON
THREE SISTERS LIGHTHOUSES
CAPE COD NATIONAL SEASHORE
NAUSET LIGHT BEACH MASS.
ANDERSON
12/12/83
SCALE OF FEET

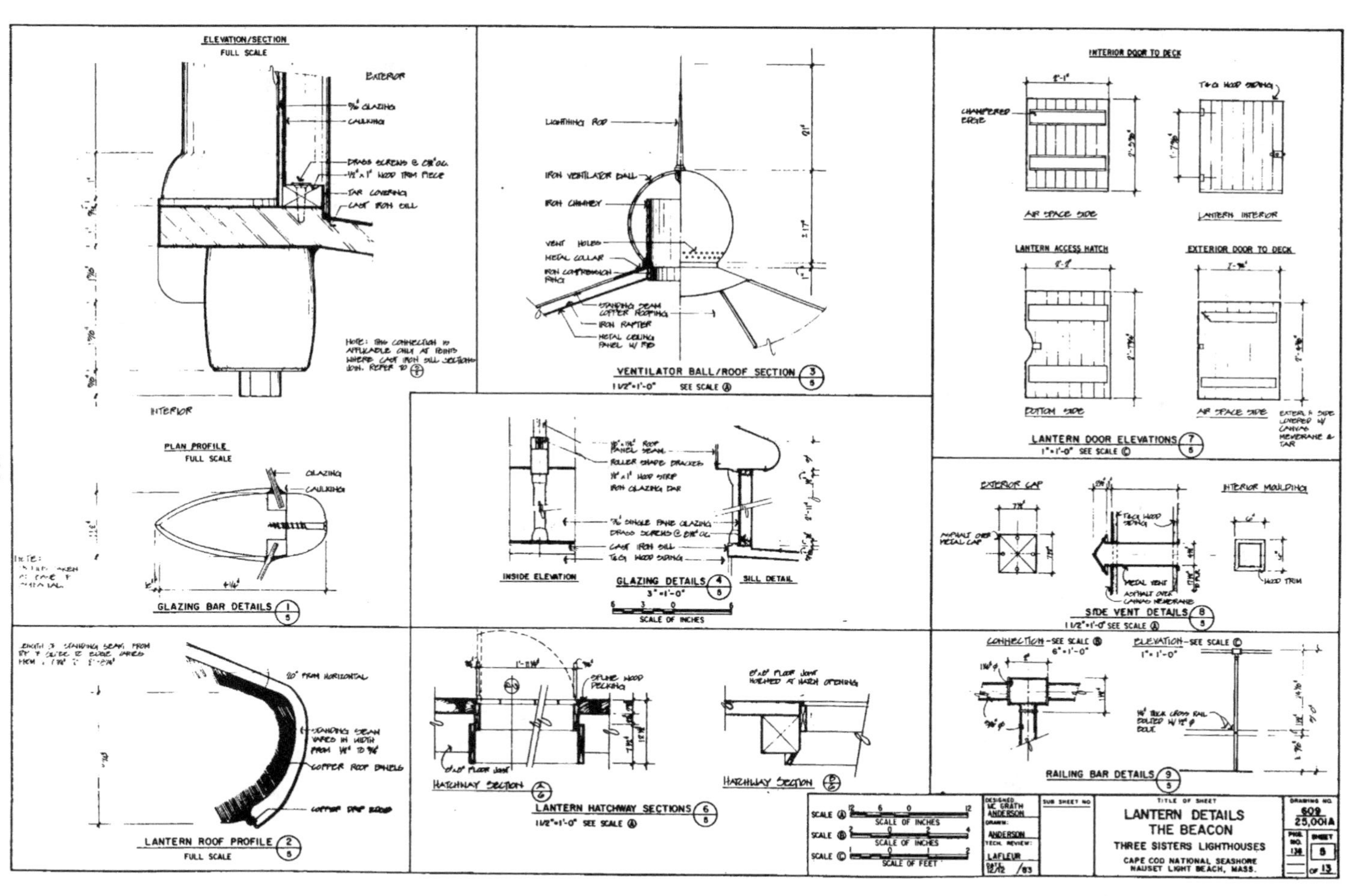
ELEVATION/SECTION
FULL SCALE
EXTERIOR
INTERIOR
PLAN PROFILE
FULL SCALE
GLAZING
CAULKING
GLAZING BAR DETAILS 1/5
VENTILATOR BALL/ROOF SECTION 3/5
1 1/2"=1'-0" SEE SCALE Ⓐ
LIGHTNING ROD
IRON VENTILATOR BALL
IRON CHIMNEY
VENT HOLES
METAL COLLAR
IRON RAFTER
INSIDE ELEVATION
GLAZING DETAILS 4/5
3"=1'-0"
SILL DETAIL
SCALE OF INCHES
INTERIOR DOOR TO DECK
CHAMFERED EDGE
AIR SPACE SIDE
LANTERN INTERIOR
LANTERN ACCESS HATCH
EXTERIOR DOOR TO DECK
BOTTOM SIDE
AIR SPACE SIDE
LANTERN DOOR ELEVATIONS 7/5
1"=1'-0" SEE SCALE Ⓒ
EXTERIOR CAP
INTERIOR MOULDING
WOOD TRIM
SIDE VENT DETAILS 8/5
1 1/2"=1'-0" SEE SCALE Ⓐ
CONNECTION - SEE SCALE Ⓑ
6"=1'-0"
ELEVATION - SEE SCALE Ⓒ
1"=1'-0"
RAILING BAR DETAILS 9/5
20° FROM HORIZONTAL
COPPER ROOF PANELS
LANTERN ROOF PROFILE 2/5
FULL SCALE
HATCHWAY SECTION
LANTERN HATCHWAY SECTIONS 6/5
1 1/2"=1'-0" SEE SCALE Ⓐ
SCALE Ⓐ SCALE OF INCHES
SCALE Ⓑ SCALE OF INCHES
SCALE Ⓒ SCALE OF FEET
DESIGNED: MC GRATH ANDERSON
DRAWN: ANDERSON
TECH. REVIEW: LAFLEUR
DATE 12/12 /83
SUB SHEET NO
TITLE OF SHEET
LANTERN DETAILS
THE BEACON
THREE SISTERS LIGHTHOUSES
CAPE COD NATIONAL SEASHORE
NAUSET LIGHT BEACH, MASS.
DRAWING NO. 609 25,001A
PKG. NO. 138
SHEET 5 OF 13

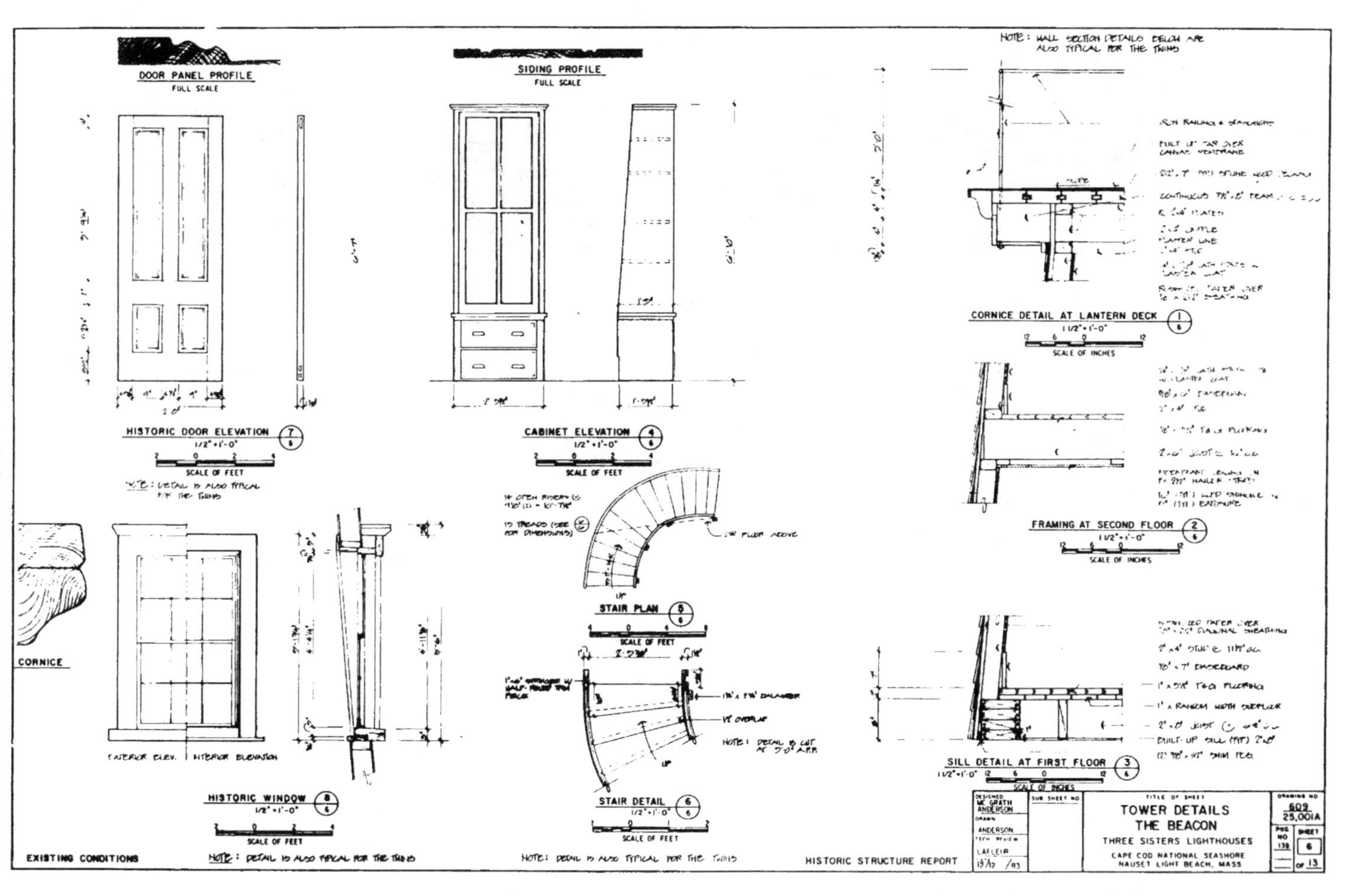
DOOR PANEL PROFILE
FULL SCALE
SIDING PROFILE
FULL SCALE
HISTORIC DOOR ELEVATION
1/2" = 1'-0"
SCALE OF FEET
CABINET ELEVATION
1/2" = 1'-0"
SCALE OF FEET
CORNICE
EXTERIOR ELEV.
INTERIOR ELEVATION
HISTORIC WINDOW
1/2" = 1'-0"
SCALE OF FEET
STAIR PLAN
SCALE OF FEET
STAIR DETAIL
1/2" = 1'-0"
SCALE OF FEET
CORNICE DETAIL AT LANTERN DECK
1 1/2" = 1'-0"
SCALE OF INCHES
FRAMING AT SECOND FLOOR
1 1/2" = 1'-0"
SCALE OF INCHES
SILL DETAIL AT FIRST FLOOR
1 1/2" = 1'-0"
SCALE OF INCHES
EXISTING CONDITIONS
HISTORIC STRUCTURE REPORT
TOWER DETAILS
THE BEACON
THREE SISTERS LIGHTHOUSES
CAPE COD NATIONAL SEASHORE
NAUSET LIGHT BEACH, MASS
609
25,001A
6
13

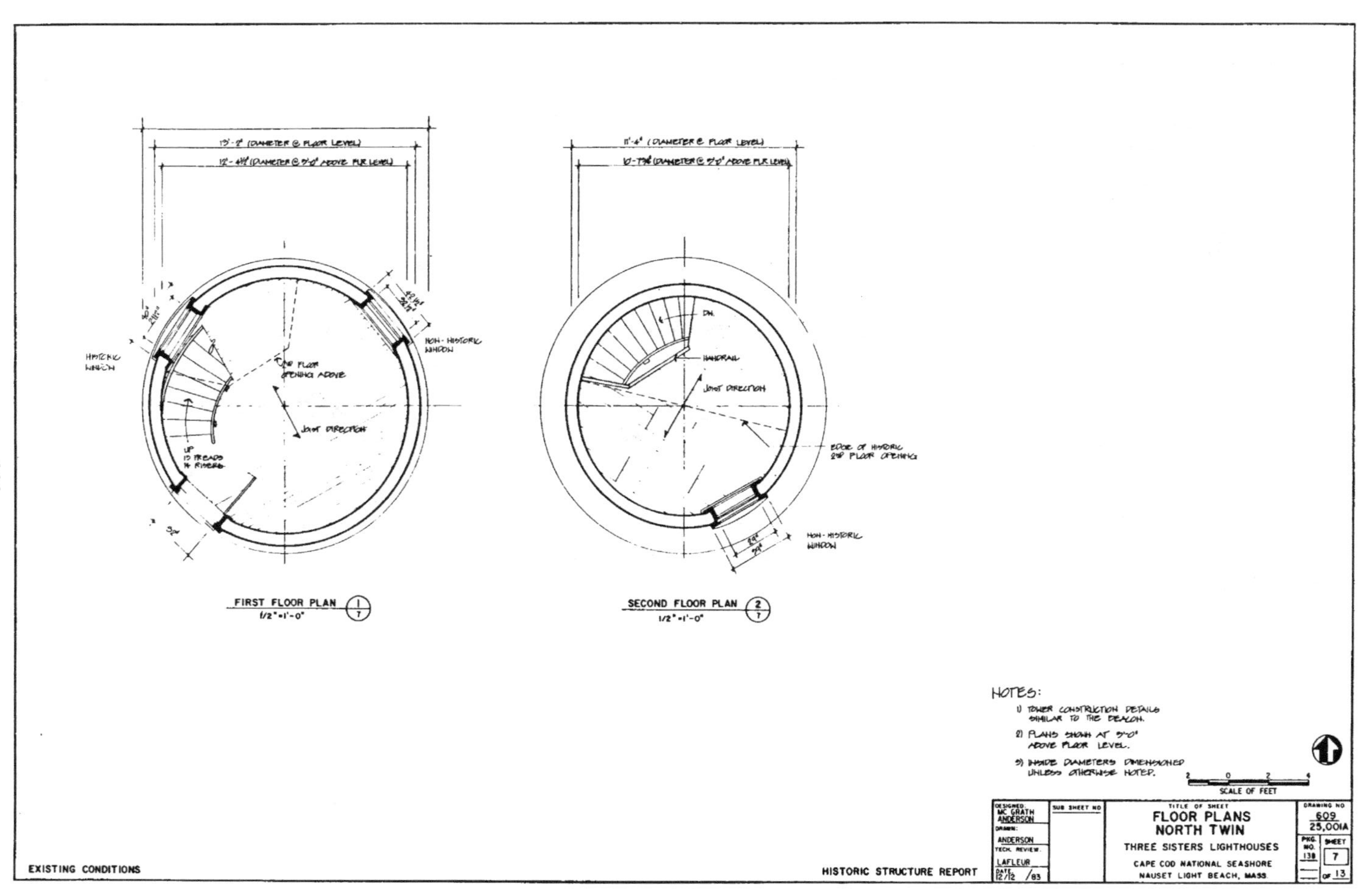
13'-2" (DIAMETER @ FLOOR LEVEL)
12'-4½" (DIAMETER @ 5'-0" ABOVE FLR LEVEL)
11'-4" (DIAMETER @ FLOOR LEVEL)
10'-7⅜" (DIAMETER @ 5'-0" ABOVE FLR LEVEL)
HISTORIC WINDOW
NON-HISTORIC WINDOW
2ND FLOOR OPENING ABOVE
JOIST DIRECTION
UP 13 TREADS 14 RISERS
DN.
HANDRAIL
EDGE OF HISTORIC 2ND FLOOR OPENING
FIRST FLOOR PLAN 1/7
1/2" = 1'-0"
SECOND FLOOR PLAN 2/7
1/2" = 1'-0"
NOTES:
1) TOWER CONSTRUCTION DETAILS SIMILAR TO THE BEACON.
2) PLANS SHOWN AT 5'-0" ABOVE FLOOR LEVEL.
3) INSIDE DIAMETERS DIMENSIONED UNLESS OTHERWISE NOTED.
SCALE OF FEET
DESIGNED: MC GRATH ANDERSON
DRAWN: ANDERSON
TECH. REVIEW: LAFLEUR
DATE: 12/12 /83
SUB SHEET NO
TITLE OF SHEET
FLOOR PLANS
NORTH TWIN
THREE SISTERS LIGHTHOUSES
CAPE COD NATIONAL SEASHORE
NAUSET LIGHT BEACH, MASS
DRAWING NO 609 25,001A
PKG. NO. 138
SHEET 7 OF 13
EXISTING CONDITIONS
HISTORIC STRUCTURE REPORT

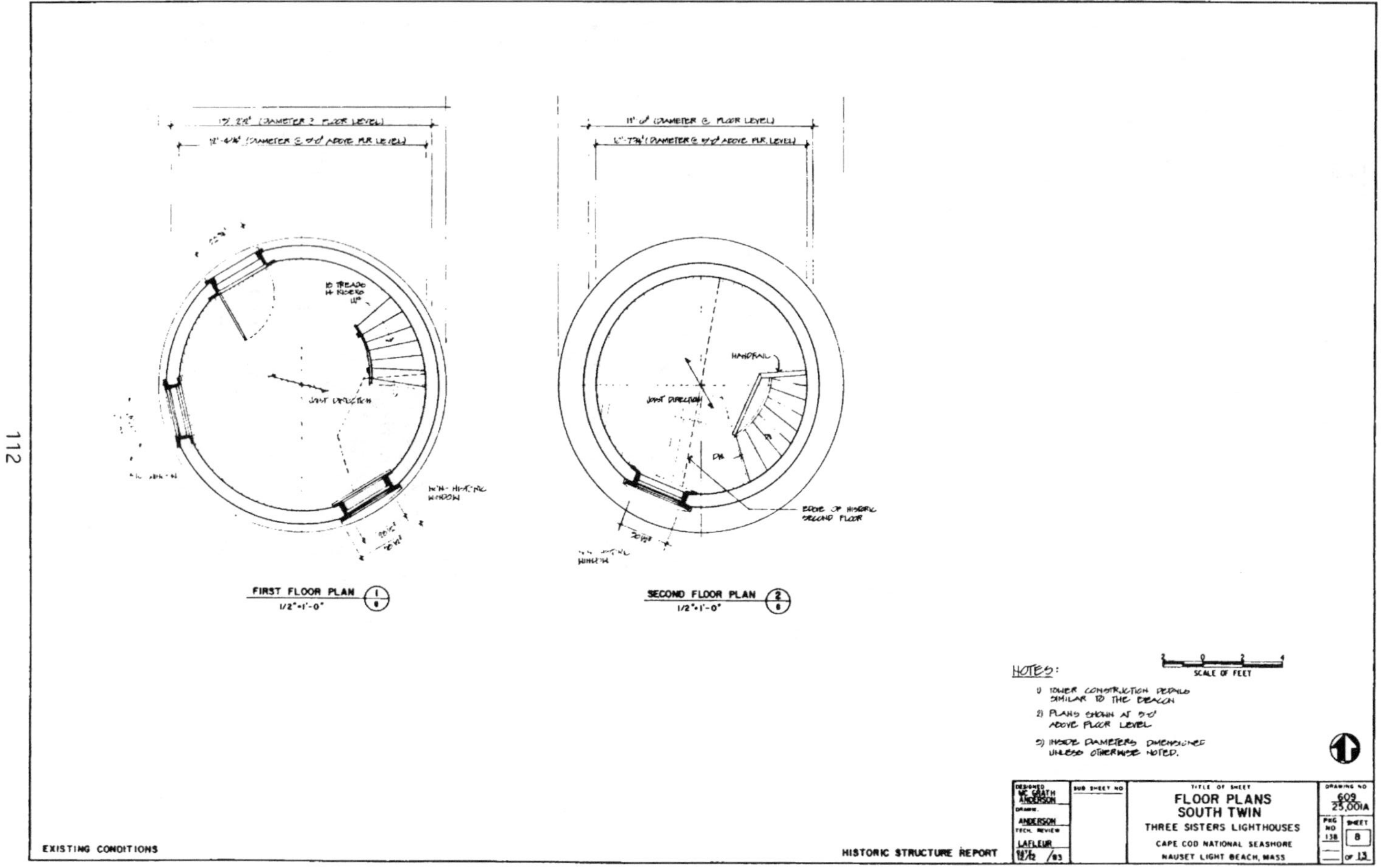
12'-4 1/4" (DIAMETER @ 5'-0" ABOVE FLR LEVEL)
11' 0" (DIAMETER @ FLOOR LEVEL)
10 TREADS
11 RISERS
UP
JOIST DIRECTION
WINDOW
HANDRAIL
EDGE OF HISTORIC SECOND FLOOR
FIRST FLOOR PLAN
1/2"=1'-0"
SECOND FLOOR PLAN
1/2"=1'-0"
NOTES:
1) TOWER CONSTRUCTION DETAILS SIMILAR TO THE BEACON
2) PLANS SHOWN AT 5'-0" ABOVE FLOOR LEVEL
3) INSIDE DIAMETERS DIMENSIONED UNLESS OTHERWISE NOTED.
SCALE OF FEET
DESIGNED MC GRATH ANDERSON
DRAWN ANDERSON
TECH. REVIEW LAFLEUR
DATE 12/12 /83
SUB SHEET NO
TITLE OF SHEET
FLOOR PLANS
SOUTH TWIN
THREE SISTERS LIGHTHOUSES
CAPE COD NATIONAL SEASHORE
NAUSET LIGHT BEACH, MASS
DRAWING NO
609
25,001A
PKG NO 138
SHEET 8 OF 13
EXISTING CONDITIONS
HISTORIC STRUCTURE REPORT

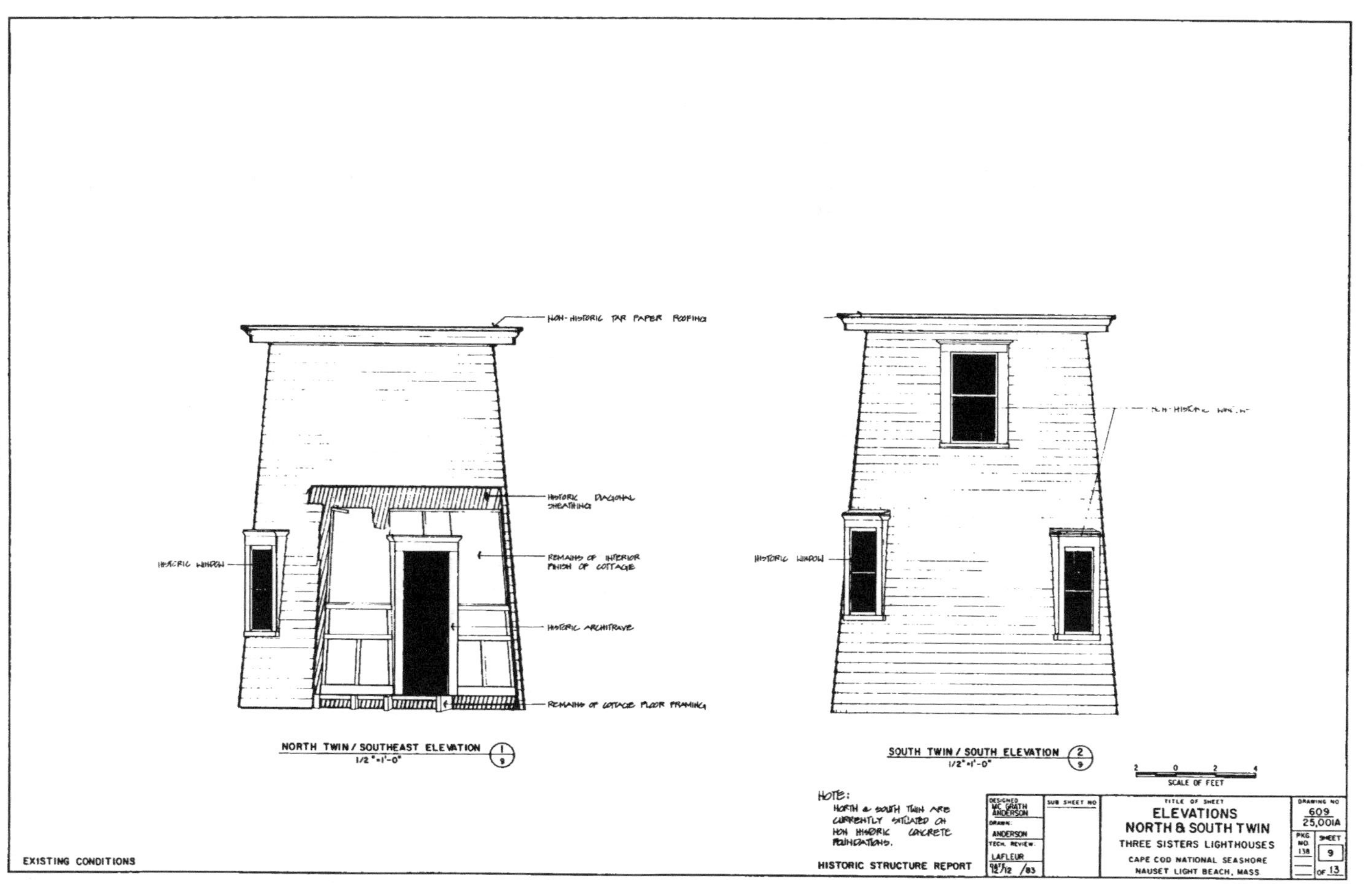

NON-HISTORIC TAR PAPER ROOFING
HISTORIC DIAGONAL SHEATHING
HISTORIC WINDOW
REMAINS OF INTERIOR FINISH OF COTTAGE
HISTORIC ARCHITRAVE
REMAINS OF COTTAGE FLOOR FRAMING
NORTH TWIN / SOUTHEAST ELEVATION
1/2"=1'-0"
NON-HISTORIC WINDOW
HISTORIC WINDOW
SOUTH TWIN / SOUTH ELEVATION
1/2"=1'-0"
2 0 2 4
SCALE OF FEET
NOTE:
NORTH & SOUTH TWIN ARE CURRENTLY SITUATED ON NON HISTORIC CONCRETE FOUNDATIONS.
HISTORIC STRUCTURE REPORT
DESIGNED MC GRATH ANDERSON
DRAWN: ANDERSON
TECH. REVIEW: LAFLEUR
DATE 12/12 /83
SUB SHEET NO
TITLE OF SHEET
ELEVATIONS
NORTH & SOUTH TWIN
THREE SISTERS LIGHTHOUSES
CAPE COD NATIONAL SEASHORE
NAUSET LIGHT BEACH, MASS
DRAWING NO
609
25,001A
PKG NO. 138
SHEET
9
OF 13
EXISTING CONDITIONS

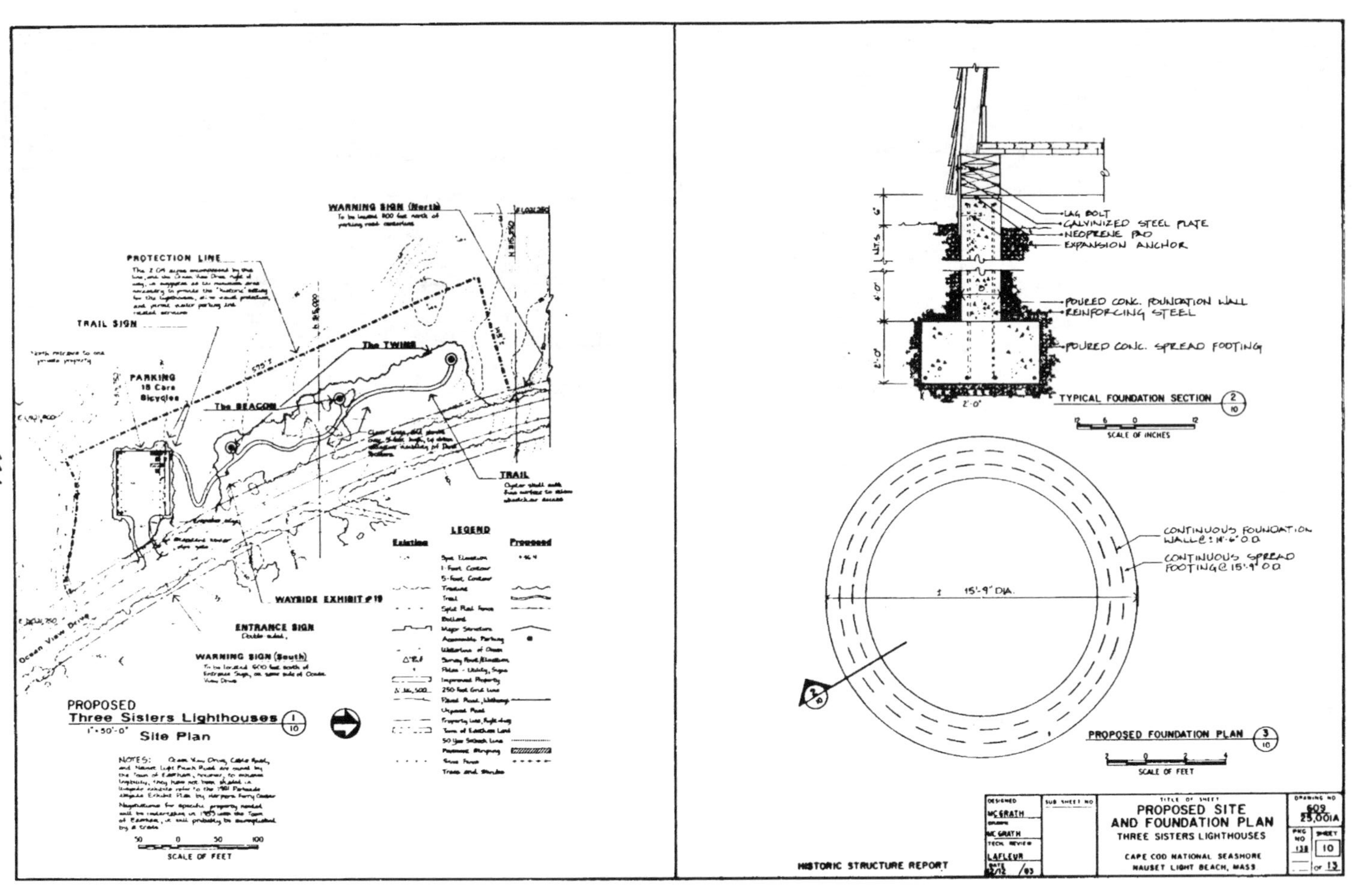
WARNING SIGN (North)
PROTECTION LINE
TRAIL SIGN
PARKING
18 Cars
Bicycles
The TWINS
The BEACON
TRAIL
LEGEND
Existing
Proposed
Spot Elevation
1-Foot Contour
5-Foot Contour
Treeline
Trail
Split Rail Fence
Bollard
Major Structure
Accessible Parking
Waterline of Ocean
Survey Point/Elevation
Poles - Utility, Signs
Improved Property
250 Foot Grid Line
Paved Road, Walkway
Unpaved Road
Property Line, Right-of-way
Town of Eastham Land
50 Year Setback Line
Pavement Striping
Snow Fence
Trees and Shrubs
WAYSIDE EXHIBIT #19
ENTRANCE SIGN
Double sided
WARNING SIGN (South)
Ocean View Drive
PROPOSED
Three Sisters Lighthouses
1"=50'-0"
Site Plan
1
10
NOTES:
50 0 50 100
SCALE OF FEET
LAG BOLT
GALVINIZED STEEL PLATE
NEOPRENE PAD
EXPANSION ANCHOR
POURED CONC. FOUNDATION WALL
REINFORCING STEEL
POURED CONC. SPREAD FOOTING
6"
4'-0"
2'-0"
TYPICAL FOUNDATION SECTION
2
10
12 6 0 12
SCALE OF INCHES
CONTINUOUS FOUNDATION WALL@14'-6" O.D.
CONTINUOUS SPREAD FOOTING@15'-9" O.D.
15'-9" DIA.
PROPOSED FOUNDATION PLAN
3
10
2 0 2 4
SCALE OF FEET
HISTORIC STRUCTURE REPORT
DESIGNED
MCGRATH
MCGRATH
TECH. REVIEW
LAFLEUR
/83
SUB SHEET NO
TITLE OF SHEET
PROPOSED SITE
AND FOUNDATION PLAN
THREE SISTERS LIGHTHOUSES
CAPE COD NATIONAL SEASHORE
NAUSET LIGHT BEACH, MASS
DRAWING NO
609
25,001A
PKG NO
138
SHEET
10
OF 13

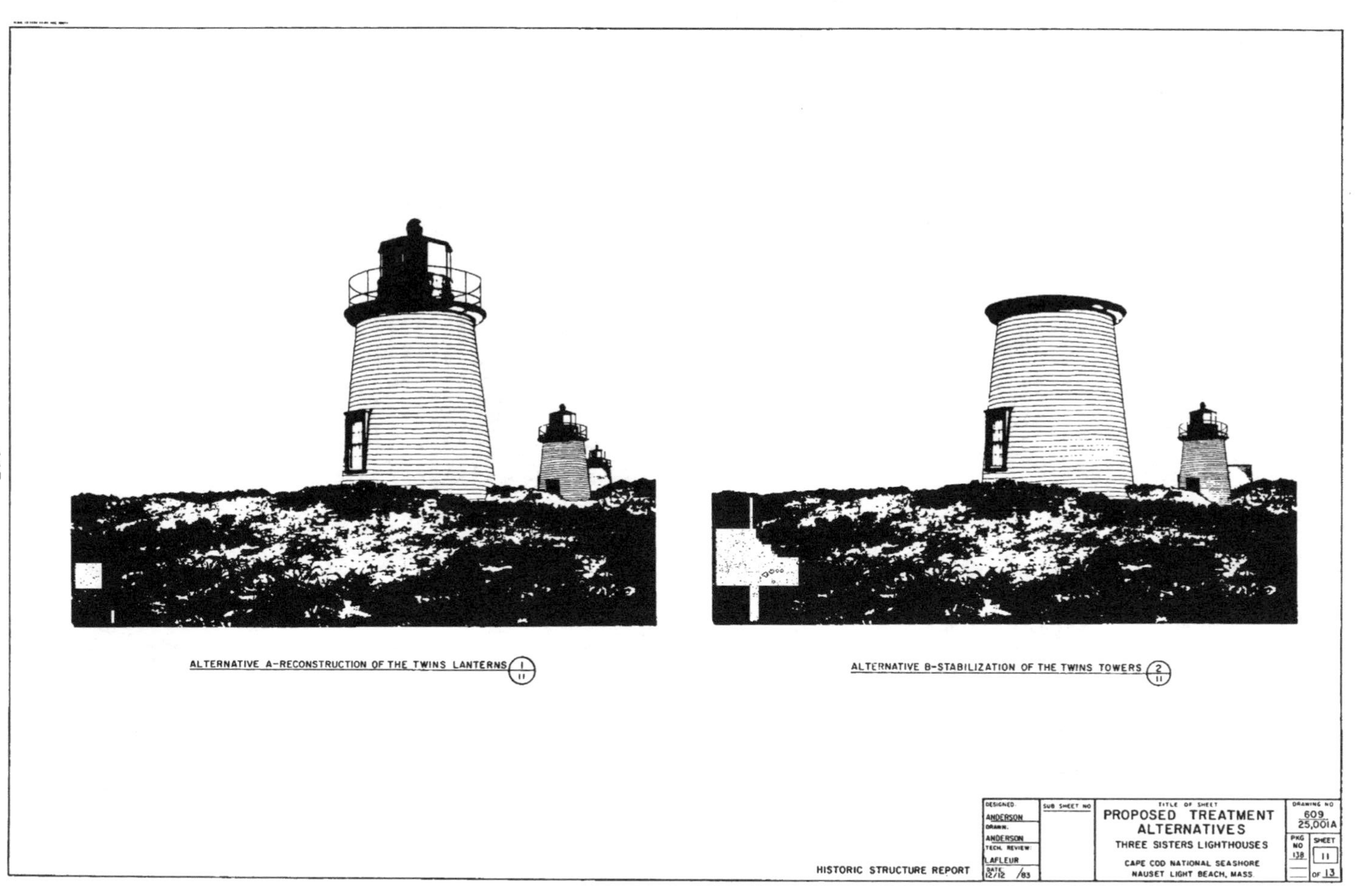
ALTERNATIVE A-RECONSTRUCTION OF THE TWINS LANTERNS
1
11
ALTERNATIVE B-STABILIZATION OF THE TWINS TOWERS
2
11
HISTORIC STRUCTURE REPORT
DESIGNED
ANDERSON
DRAWN:
ANDERSON
TECH. REVIEW:
LAFLEUR
DATE
12/12 /83
SUB SHEET NO
TITLE OF SHEET
PROPOSED TREATMENT
ALTERNATIVES
THREE SISTERS LIGHTHOUSES
CAPE COD NATIONAL SEASHORE
NAUSET LIGHT BEACH, MASS.
DRAWING NO
609
25,001A
PKG NO
138
SHEET
11
OF 13

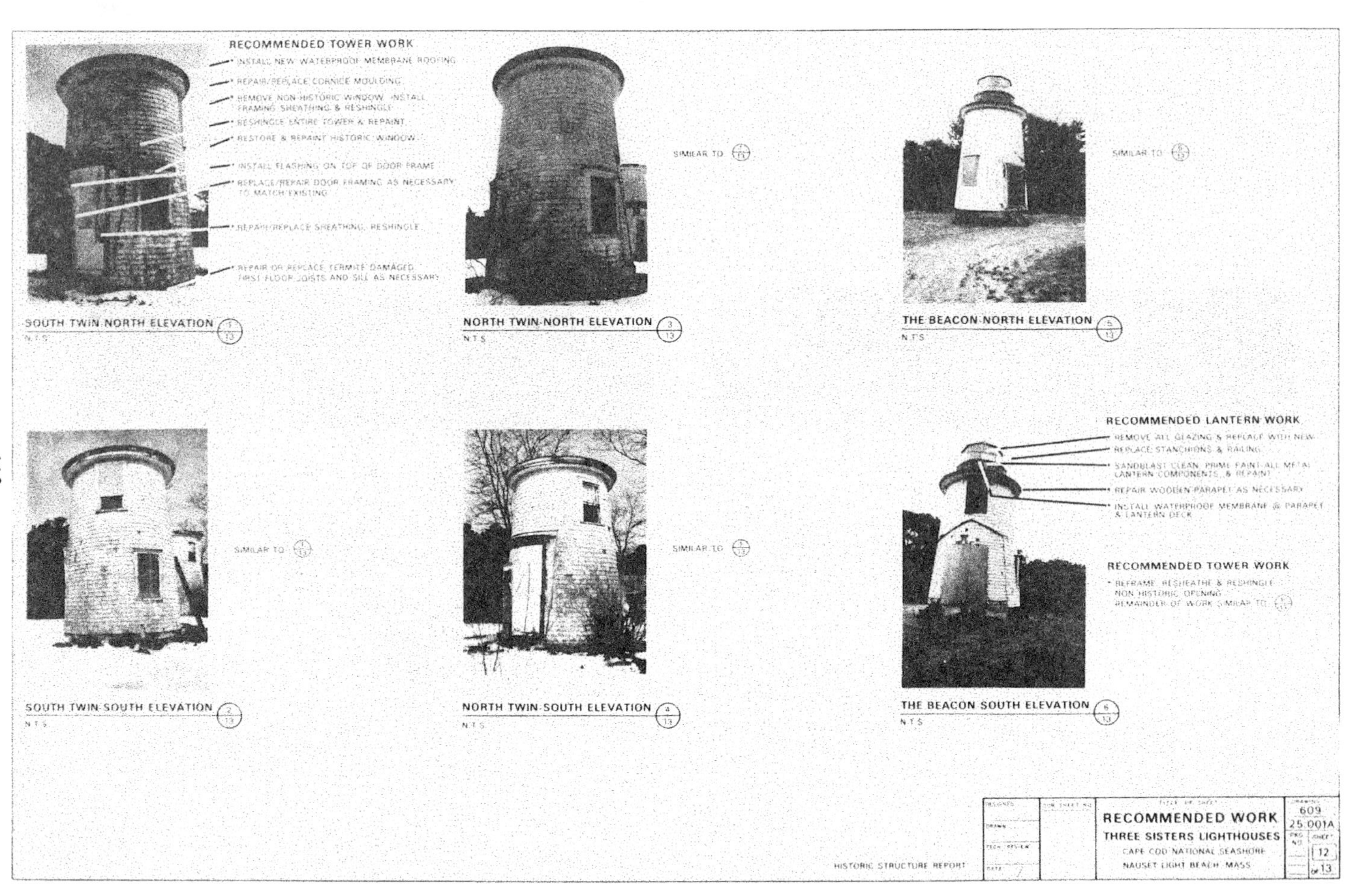
RECOMMENDED TOWER WORK
INSTALL NEW WATERPROOF MEMBRANE ROOFING
REPAIR/REPLACE CORNICE MOULDING
REMOVE NON-HISTORIC WINDOW, INSTALL FRAMING SHEATHING & RESHINGLE
RESHINGLE ENTIRE TOWER & REPAINT
RESTORE & REPAINT HISTORIC WINDOW
INSTALL FLASHING ON TOP OF DOOR FRAME
REPLACE/REPAIR DOOR FRAMING AS NECESSARY TO MATCH EXISTING
REPAIR/REPLACE SHEATHING, RESHINGLE
REPAIR OR REPLACE TERMITE DAMAGED FIRST FLOOR JOISTS AND SILL AS NECESSARY
SOUTH TWIN NORTH ELEVATION 1/13
N.T.S.
NORTH TWIN NORTH ELEVATION 3/13
N.T.S.
SIMILAR TO
THE BEACON NORTH ELEVATION 5/13
N.T.S.
SIMILAR TO
SOUTH TWIN SOUTH ELEVATION 2/13
N.T.S.
SIMILAR TO
NORTH TWIN SOUTH ELEVATION 4/13
N.T.S.
SIMILAR TO
RECOMMENDED LANTERN WORK
REMOVE ALL GLAZING & REPLACE WITH NEW
REPLACE STANCHIONS & RAILING
SANDBLAST CLEAN, PRIME PAINT ALL METAL LANTERN COMPONENTS, & REPAINT
REPAIR WOODEN PARAPET AS NECESSARY
INSTALL WATERPROOF MEMBRANE @ PARAPET & LANTERN DECK
RECOMMENDED TOWER WORK
REFRAME, RESHEATHE & RESHINGLE NON HISTORIC OPENING
REMAINDER OF WORK SIMILAR TO
THE BEACON SOUTH ELEVATION 6/13
N.T.S.
HISTORIC STRUCTURE REPORT
RECOMMENDED WORK
THREE SISTERS LIGHTHOUSES
CAPE COD NATIONAL SEASHORE
NAUSET LIGHT BEACH, MASS.
609
25.001A
12
of 13

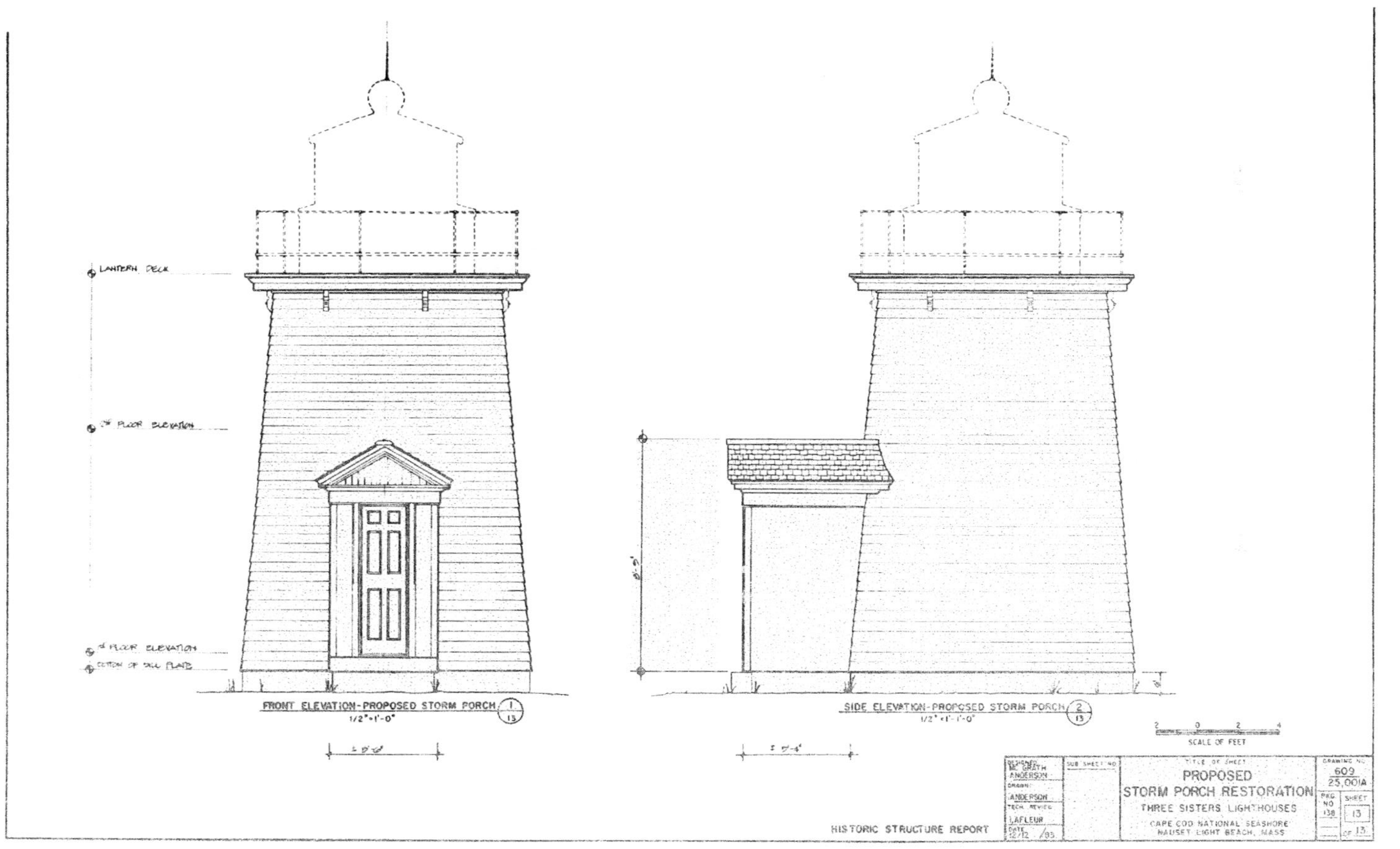

LANTERN DECK
FRONT ELEVATION-PROPOSED STORM PORCH 1/13
1/2"=1'-0"
SIDE ELEVATION-PROPOSED STORM PORCH 2/13
1/2"=1'-0"
2 0 2 4
SCALE OF FEET
HISTORIC STRUCTURE REPORT
DESIGNED MC GRATH ANDERSON
DRAWN ANDERSON
TECH. REVIEW LAFLEUR
DATE 12/12/95
SUB SHEET NO.
TITLE OF SHEET
PROPOSED
STORM PORCH RESTORATION
THREE SISTERS LIGHTHOUSES
CAPE COD NATIONAL SEASHORE
NAUSET LIGHT BEACH, MASS
DRAWING NO.
609
25,001A
PKG NO 138
SHEET 13
OF 13

APPENDIX B

The Armeria's sister ship the Azalia

sister ship the <u>Azalia</u>

THE AZALEA.

GENERAL DESCRIPTION.

There will be one right-handed cast-iron screw-propeller, of about 8 feet diameter and suitable pitch, driven by an inverted-cylinder surface-condensing two-cylinder fore-and-aft compound engine, the cylinders to be 19 inches and 36 inches in diameter, with a stroke of 28 inches.

BOILER.

There is to be one cylindrical boiler, 11 feet 9 inches diameter outside of the smallest course of shell, and 11 feet 4 inches long from outside to outside of ends.

There are to be two Fox's patent corrugated furnaces, 40 inches in diameter; inside corrugations to extend from the front end to the back tube-head and secured thereto by means of a single seam of rivets of diameter and pitch to suit the thickness of the plates.

The back-connection will be 29½ inches long at the bottom and 26 inches at the top, increased in a direction parallel with the length of the boiler, and the side sheets are to be carried eccentric to the shell, so that the distance between the shell and said side-sheet will be about 5½ inches in the clear at the bottom, increase upward to a distance, in the clear, of about 7½ inches, increased diametrically at the height of the top row of the tubes. The crown-sheet is to be horizontal and stayed by means of wrought-iron girders, properly spaced, and formed at the ends to have bearings on the front, back, and crown plates, and have a clear space of not less than 1½ inches between the crown-sheet and the bottom of the girders.

The shell of the boiler will be made in two lengths of shell, longitudinally, the smaller course toward the front, or firing end, and suitable thickness to meet the requirements of the United States steam-boiler inspection for a working pressure of 110 pounds per gauge. The circumferential seam will be a double lap-riveted seam, and the longitudinal seams are to be properly spaced to clear all fittings, and are to be riveted and strapped to suit the requirements of the United States steam-boiler inspection for a working pressure as stated above. The end plates shall be made in not more than two plates in each end plate, the seam joining the plates to be horizontal, and situated at a proper height above the tops of the tubes; seams to be double-riveted zigzag, with the calking edge upon the outside. There are to be one hundred and sixty-six 3¼ inches outside diameter American lap-welded charcoal-iron boiler-tubes; the distance between the tube-sheets, from inside to inside, to be about 8 feet, spaced not less than 4½ inches horizontally and vertically; a vertical space of about 7½ inches, in the clear, to be left between the tubes in the center of the boiler. Stay-tubes to be put in in a sufficient number, and to the satisfaction of the representative of the Light-House Board. The stay-tubes are to be proportioned at the ends for cutting suitable threads; the back-end thread to be minus and the front-end thread plus, so that the tube will pass freely through the front head. Each stay-tube to be fitted with a flat-cheek hexagonal nut, the other tubes to be expanded and beaded in the very best manner. Tubes, etc., to be arranged according to working drawing furnished by the Light-House Board.

The scantlings of the boiler to be as follows: Shell-plate, $\frac{13}{16}$ inch in thickness; front and rear tube-head, $\frac{3}{4}$ inch; back-head of back-connection, $\frac{1}{2}$ inch; side sheets of back-connection, $\frac{7}{16}$ inch; crown-sheet of back-connection, $\frac{9}{16}$ inch; upper part of end plates, $\frac{9}{16}$ inch; furnace, $\frac{1}{2}$ inch. The back-connection will be stayed to the shell in a substantial manner, and in accordance with the rules of the Board of Supervising Inspectors governing the construction of marine boilers to be used in salt-water. The strain on the stays shall not exceed 5,000 pounds per square inch of section at the bottom of the thread. The front and back ends will be provided with stays extending from outside to outside of shell, with hexagonal-nut washers outside, and nuts and castings inside to stiffen the end plates properly. A pair of 3 by 3 by $\frac{7}{16}$ inches angle-iron are to be securely riveted to the inside of the end plates, one on each side of the stay, and about $\frac{3}{4}$ of an inch farther apart than the diameter of the stay. The angle-irons are to extend horizontally across the ends a sufficient distance beyond the outside stays to insure good work. The stays are to be spaced far enough apart to admit a man to pass between them to inspect the boiler. These stays are to be proportioned to stand not more than 6,000 pounds per square inch of section at the root of the thread.

BOILER-FITTINGS.

An independent stop and safety-valve will be fitted to the boiler; the safety-valve is to be on the shell of the boiler at the up-take end, and an exhaust-pipe of copper, with a properly designed end of ornamental pattern, and branch-pipe to lead steam up and out, properly attached in a substantial manner, and carried to such height as may be directed. The safety-valve is to be of such spring safety-valve pattern as is suitable for marine purposes, and of such size as is required by law. The stop-valves are to be of an approved form, and will be attached to the end of the boiler nearest the engine; they are to be furnished complete with hand-wheels, stuffing-box, and bonnet. A 5-inch diameter wrought-iron pipe will extend from this valve inward along the top of the boiler, one end being made so as to prevent steam from entering, and the other end screwed into the valve-body. The pipe will be properly fastened to the inside of the boiler, about 3 inches from the upper side; two rows of $\frac{1}{4}$-inch diameter holes shall be drilled through the top side of said pipe, so that the sum of the area of said holes shall be 50 per cent. in excess of the area of the pipe; 11 by 15 inch elliptical man-holes and 5 by 7 inch elliptical hand-holes are to be cut through the boiler, in such position and number as may be directed, and these openings are to be properly re-enforced and fitted complete, with good hinges between the plates and boiler to insure tight joints and strength. The plates, bolts, and bridge-bars are to be of the best quality of material and workmanship.

The furnaces are to be fitted complete with bridge-walls, grate-bars, bearing-bars, front castings, firing and ash-pit doors, in a substantial and complete manner; the bridge-walls to be furnished with soapstone shapes to suit the diameter of furnace. The boiler is to be furnished complete with bottom and surface blow-valves and pipe, feed-pipe and valves, foam-pipe, glass and cock gauges, 6-inch dial-gauges, and all fittings necessary to place them in complete working order, such as sea-connections, firing-tools, and all necessary fittings. The boilers are to be securely strapped to the cradles and chocked in a substantial and neat manner, to prevent any movement endwise or athwartship on the cradles. The smoke-box will be constructed as shown on the drawing, and will be built double up to within 4 inches of the top of sheet-iron casing fitted to deck; the passages are not to be reduced in area to less than one-fifth of the grate area at any point. The front of the smoke-box will slope outward, and will be framed so that the openings for doors will admit of the tubes being removed without deranging any of the structure.

These doors are to be hinged at the top edge and swing vertically, and are to be finished complete, with hinges, clamps, eye-bolts, baffle-plates, and lifting-chains, with pulley-leaders and lines leading to some accessible and convenient point. The material of smoke-box and funnel is to be of sufficient strength for first-class work, and must be finished complete, in place, with bearings for properly supporting the weight, stays, stay-band, umbrellas, drip and shield plates, and all fittings necessary to complete the work, to meet the requirements of the United States inspector of steam-vessels. All valves to be of the best material and workmanship, properly attached and fitted in place. The engine and boiler rooms are to be fitted complete with wrought-iron platforms, ladders, and floor-plates, as shown on the drawings, or as may be directed during construction, to render every part of engine and boilers accessible and convenient.

When the boiler has been properly tested, etc., asbestos and felt, or other approved covering, shall be put on by the contractor, to the satisfaction of the representatives of the Light-House Board.

MATERIAL.

All the material of the boilers is to be Siemens-Martin mild-steel, excepting the tubes, with a tensile strength of not under 60,000 pounds nor more than 65,000 pounds per square inch of original cross-section, and an elongation of 23 per cent. in 8 inches. Rivets are to be of the very best quality of steel, of 50,000 to 55,000 pounds tensile strength per square inch, and an elongation of 30 per cent. in 8 inches.

TRIAL-TRIP.

In addition to the necessary trials of the machinery at the dock, a trial-trip is also to be made, of about twelve hours' duration, or as may be directed by the Light-House Board, at the expense of the contractor, and the engine must develop the required 400 indicated horse-power when the engine is making 107 revolutions per minute, with a coal-consumption as hereinbefore specified, and a cut-off in the high-pressure cylinder of one-half the stroke, links in full-gearing and steam, per gauge, at 100 pounds pressure per square inch.

All bearings, journals, crank-pins, and other parts of the engine to show no tendency to heat or grip, but to run smoothly, the engine to pass its centers without shock or noise. The machinery must work on this trial-trip to the entire satisfaction of the representative of the Light-House Board, and if any defects should develop on the trial, subsequent trials, at the expense of the contractor, as described above, will be made until every part of the machinery has been proven to be in accordance with the requirements of these specifications, to the satisfaction of the Light-House Board.

The contract for the *Azalea* requires the delivery of this vessel at the light-house depot Tompkinsville, Staten Island, New York, on or before seven calendar months from the date of the approval of the contract by the Secretary of the Treasury. This contract was approved on February 28, 1890, so her delivery may be expected on or about September 28, 1890.

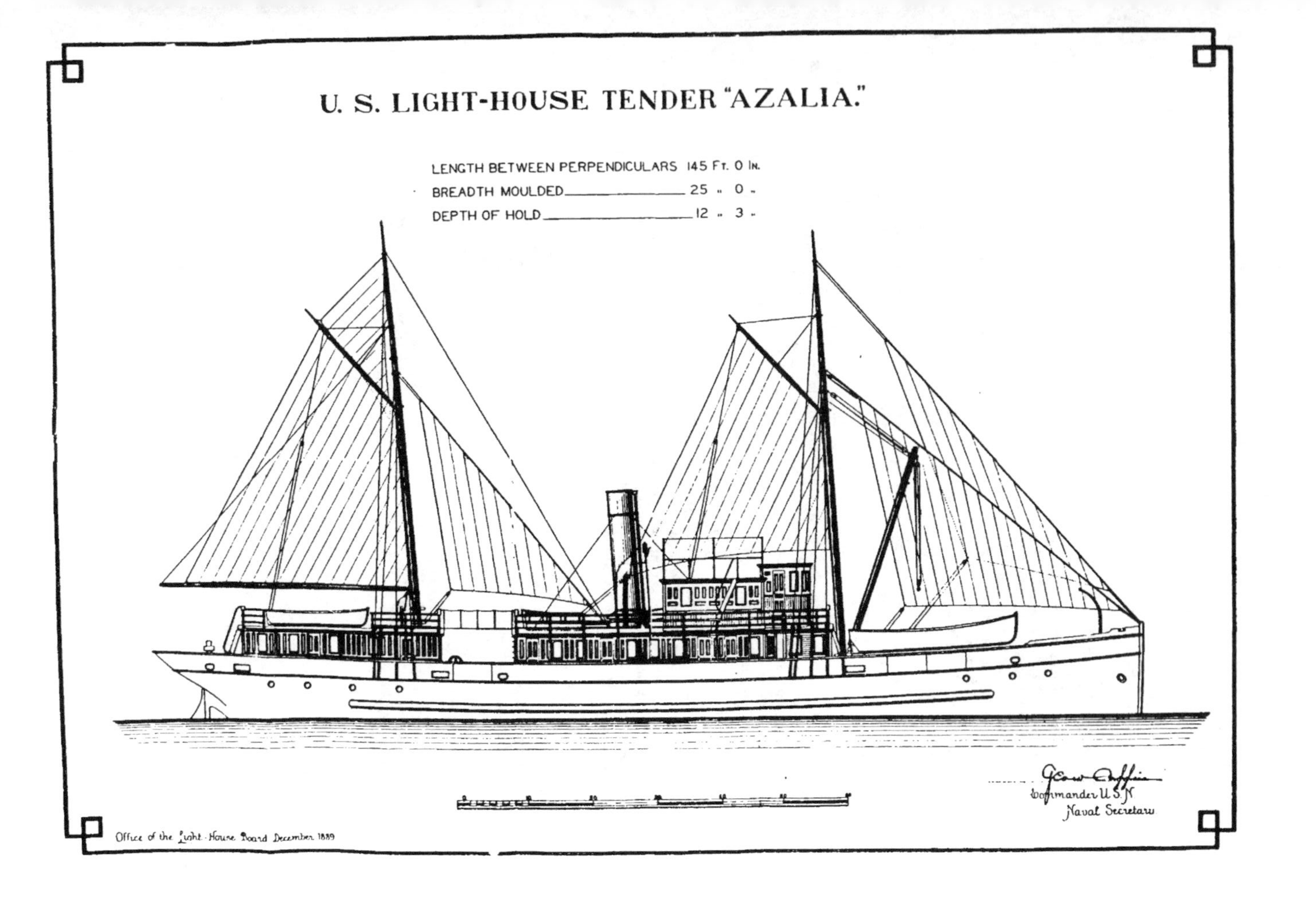
U. S. LIGHT-HOUSE TENDER "AZALIA."
LENGTH BETWEEN PERPENDICULARS 145 FT. 0 IN.
BREADTH MOULDED 25 " 0 "
DEPTH OF HOLD 12 " 3 "
Commander U.S.N.
Naval Secretary
Office of the Light-House Board December 1889

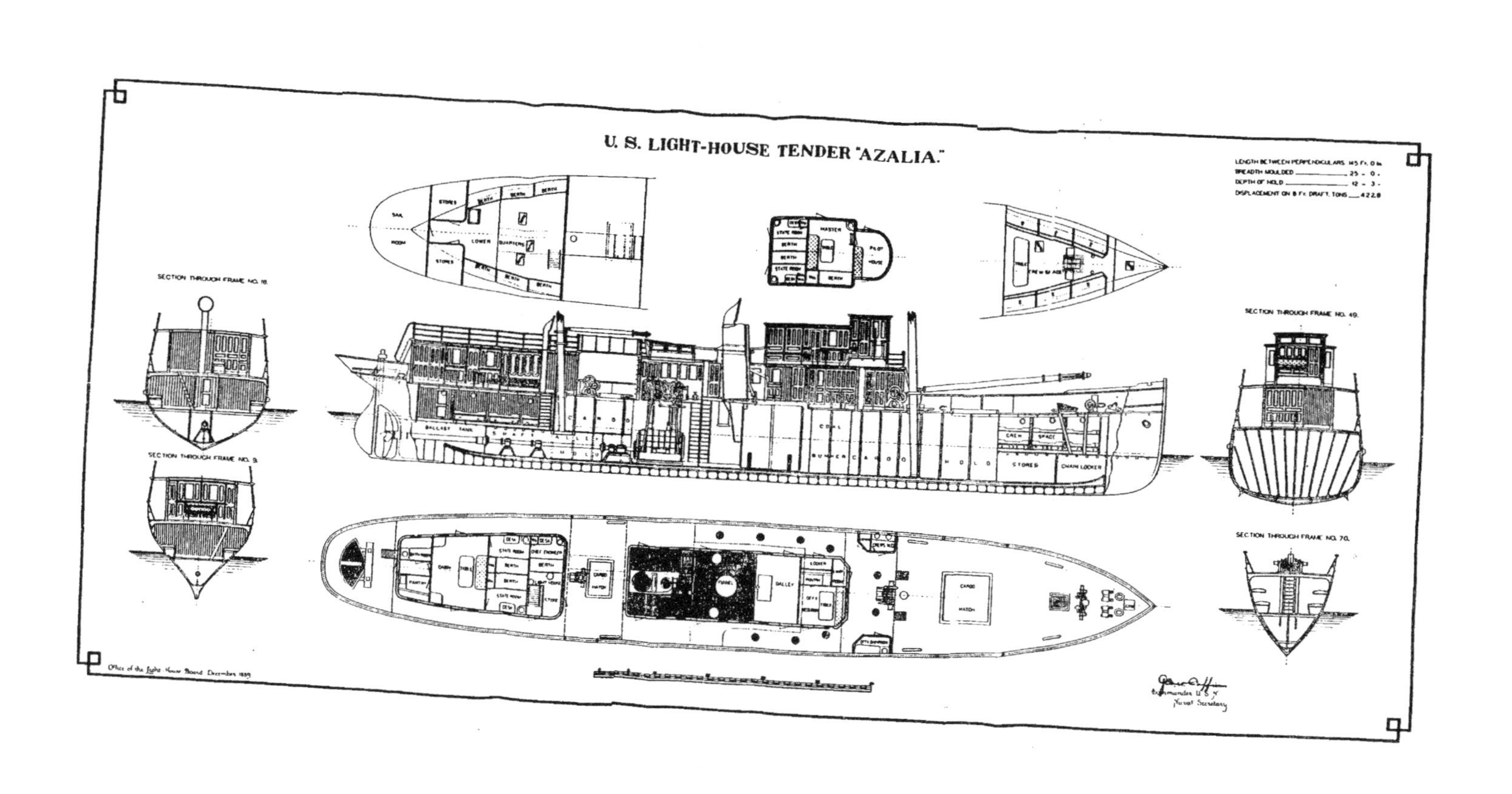

U. S. LIGHT-HOUSE TENDER "AZALIA."
LENGTH BETWEEN PERPENDICULARS 145 Ft. 0 In.
BREADTH MOULDED 25 - 0 -
DEPTH OF HOLD 12 - 3 -
DISPLACEMENT ON 8 Ft. DRAFT, TONS 422.8
SECTION THROUGH FRAME NO. 18.
SECTION THROUGH FRAME NO. 9.
SECTION THROUGH FRAME NO. 49.
SECTION THROUGH FRAME NO. 70.
STORES
BERTH
LOWER
QUARTERS
MASTER
STATE ROOM
PILOT
HOUSE
CREW SPACE
BALLAST TANK
CARGO
COAL
BUNKER
HOLD
STORES
CHAIN LOCKER
CABIN
TABLE
PANTRY
CHIEF ENGINEER
FUNNEL
GALLEY
LOCKER
CARGO
HATCH
Office of the Light House Board December 1889
Commander U.S.N.
Naval Secretary

APPENDIX C

Regulations for the United States
Lighthouse Service
1911

INSTRUCTIONS TO LIGHT KEEPERS.

The official title of the light keepers at a station shall be keeper (employee in charge), first assistant keeper, second assistant keeper, third assistant keeper, etc.

The term "light keepers" shall include the keeper and all assistant keepers at a station.

Qualifications of keepers.—Keepers and assistant keepers must be able to read and write, and be in every respect competent to discharge the duties of keeper.

In the absence of the keeper his duties will devolve upon the senior assistant keeper present.

Weekly inspection.—Keepers will make a personal weekly inspection of the station, including quarters of all assistants and laborers. This inspection of quarters will be made in the presence of the subordinate concerned at a reasonable hour, and record made of condition in the journal.

Absences.—Light keepers shall be allowed, when practicable, 30 days' leave of absence each year, exclusive of Sundays and legal holidays, provided that, if necessary, they furnish, at their own expense, competent substitutes satisfactory to the inspector.

When a keeper is away from his station on authorized leave of absence the pay and subsistence of his substitute, in case it is necessary for him to employ one, shall be a matter of personal arrangement between the keeper and the substitute.

Keepers leaving a station temporarily must notify the senior assistant present of their departure, return, probable duration of absence and any other matter necessary to an intelligent execution of their duties.

Keepers will require assistants to obtain permission before leaving the station, to report their departure, return, and, if practicable, directions for reaching them in an emergency.

At isolated stations, with or without quarters, all absences of whatever nature will be reported to the inspector on the proper form.

All cases of sickness or other disability preventing a proper performance of duty will be reported as "absence caused by sickness, etc."

Stations with quarters: Keepers are not considered absent from duty when engaged in regular routine of station, such as getting supplies, mail, etc., and attending church, except where such work requires a trip of several hours' duration. In the latter case, an absence of over six hours in the daytime, or any absence at night, shall be reported.

Stations without quarters: A keeper at his authorized home or en route between the light structure and his home, while off watch, shall not be considered as absent from the station. Absences of any other character will be reported as prescribed in the preceding paragraph.

Special or doubtful cases not covered by these regulations will be referred to the inspector for decision.

Light stations must never be left unattended.—Either the keeper or an assistant must always be present on duty. At stations having no assistant, the keeper must arrange for some competent person to take charge during his absence, which should not extend beyond sunset, except in an emergency, at isolated stations, or unless authorized by the inspector.

In cases of sickness or other disability, light keepers must provide efficient reliefs, and when the disability is likely to be of long duration, prompt report of the fact must be made to the inspector.

Watches.—At stations having an assistant, watches must be kept and so divided that an equal share of work and desirable hours of watch shall fall to the keeper and to each assistant. Watches shall be stood in such place and manner as to give continuous and the best possible attention to the light and the fog signal when in operation.

At stations having no assistant, the keeper must visit the light at least twice between 8 p. m. and sunrise, and on stormy nights the light must be constantly looked after.

Length of time to exhibit lights.—Lights must be exhibited punctually at sunset and kept burning at full intensity until sunrise, when the light will be extinguished and the apparatus put in order without delay for relighting. When not in use, the illuminating apparatus must be covered, and the lantern curtains hung.

At stations where navigation is closed by ice in the winter, and in the absence of definite instructions from the inspector, lights may be extinguished when navigation ceases, but must always be shown when it is at all possible for vessels to benefit by them. The time of discontinuance and of relighting must be promptly reported. Keepers of isolated stations, when in doubt, should be governed in exhibiting their lights by the action of those less isolated.

All machinery must be inspected and turned over daily by hand or other power.

Tests of fog-signal engines.—Fog signals operated by oil engines, or otherwise than by steam, will be worked at full pressure at least 15 minutes each week.

Steam fog signals will be operated at full pressure at least 15 minutes each month.

The weekly and monthly tests prescribed shall be noted on the monthly fog signal reports (Forms 302 and 66) "Remarks" column. Special reports will be forwarded of any defects in the apparatus requiring action by the inspector's office.

Fog signals will be operated when the atmosphere is not clear enough to see given objects 5 miles distant, unless special instructions are given to the contrary.

Duplicate fog-signal plants.—When any part of the plant or equipment is in duplicate, each set should be given an equal amount of usage.

Cleanliness of station.—Light keepers must keep their premises clean and well whitewashed; grounds in order; all the inside painted work of the lanterns well washed, and, when required, they must retouch it with paint. The spare articles embraced in the list of allowances must be kept on hand and examined frequently, and should be kept clean and in order for use.

Property returns for light stations.—A property return shall be prepared for each light station by the keeper, showing the property on hand at his station on June 30 of each year, and the property received and expended during the fiscal year. One copy must be forwarded to the inspector not later than August 1. An inventory must be taken and the property reported in the return must be checked by actual count, and all discrepancies must be explained in a letter attached to the return, or by certificates for lost property. For use in making up the property return and examining the same, there must be kept at each station a file of all invoices or bills for property received. Account must be kept of all supplies expended, either by memorandum receipts or in a memorandum book.

The property records at the station shall from time to time be subjected to such examination at the station, to verify their correctness, as can be arranged for by the inspector, and the annual return will be verified in the district office.

Erection of buildings.—Keepers shall not erect large buildings of a permanent character on lighthouse premises, except as such work may be authorized. Keepers may erect, at their own expense, minor buildings, such as henhouses. Approval must first be obtained, however, from the lighthouse inspector, as to the location, size, exterior appearance, disposition upon the removal of the keeper, etc.

3412°—11——11

Salutes.—Vessels displaying the lighthouse pennant having on board officials or persons entitled to such courtesy, shall be saluted with three strokes of the bell, upon arrival at, departure from, or when passing a light station equipped with a bell. Bells having no inside clapper should have connecting rods detached before saluting with the striker.

Boats.—Boats shall be marked and known by their official numbers assigned by the inspector. Boats shall be used for official purposes only, which shall be construed to include all work connected directly or indirectly with the care and maintenance of the station and the personnel thereof.

Station boats must never be used for heavy freight or by working parties, except under the supervision of the keeper.

Keyboard.—When practicable, a keyboard will be kept in a safe and accessible place where all official keys must be hung by light keepers when absent from station.

The official allowance list will be strictly adhered to and no other supplies, implements, or equipment than those authorized officially shall be used in the care and maintenance of a light station.

Necessary supplies will be issued daily by the keeper in the presence of an assistant, if practicable. The exact quantities thus expended must be entered immediately in the record for entering expenditures.

The 5-gallon brass cans shall be used for the daily issue of mineral oil and must be completely emptied and cleaned before refilling. In refilling, decant all except about 1 inch of the oil from the transporting can into the brass can. The remaining inch of oil in the transporting can is to be then drawn off in a receptacle kept for the purpose and set aside for use in the hand and table lamps.

In drawing oil from the brass can for the daily consumption in the lens lamp, again lay aside the last inch of oil in the bottom of this can for use in the hand or table lamps. With oil vapor lamps, the additional precaution must be taken of passing the oil through the filter the day before and setting aside the last layer of oil in the bottom of the filter for use in the hand or table lamps.

Oil must never be siphoned from the transporting cans or any other method of emptying employed than decanting. The oil must be strained when decanted and again when poured into the lamp.

Precautions against fire.—Every precaution must be taken against fire. Fire buckets, when provided, must be kept filled with water, and other apparatus accessible, and in a fixed place.

Liquid chemical fire extinguishers must be kept charged, recharged at least once every year, in a fixed and easily accessible place, and the personnel instructed in their use.

Ashes or sand must be kept available for use on burning mineral oil, gasoline, etc.

Should oil or gasoline become ignited, the best means of extinguishing it is by the use of the liquid chemical extinguishers, being sure to direct the stream at the base of the fire. In the absence of a patent extinguisher, sand, earth, or ashes should be used. A heavy woolen blanket thrown over the fire will also tend to smother it.

Fires or lights in buildings must never be left unattended.

Smoking is prohibited in all buildings belonging to the Lighthouse Service, except in living quarters.

Open lights will never be taken into rooms used for storing oils, paints, gasoline, or other inflammable materials.

Only safety matches, or those furnished by the Lighthouse Service, will be allowed on lighthouse premises.

Oily waste, rags, and other means of spontaneous combustion will be promptly destroyed.

Uniforms.—Light keepers shall wear the uniform prescribed by the Uniform Regulations and in the manner indicated in Part I, General Instructions to all Employees, page 149.

Regulation working clothes shall be worn when engaged in dirty work.

Regulation aprons shall be worn when cleaning lamps and lenses.

Subsistence.—Each appointed light keeper is by law entitled to receive one ration per day, or in the discretion of the Commissioner, commutation therefor at the rate of 30 cents per ration. Such provisions as are contracted for by the inspector may be furnished in kind to stations where, in the opinion of the inspector, it is for the best interests of the service. Where rations in kind are delivered, the contract cost will be deducted from the allowance of 30 cents per day and the difference paid in cash as a commutation. Where no ration in kind is issued, the allowance of 30 cents per day will be paid in cash.

Efficiency and commendation.—Letters of commendation may be issued by the inspector to light keepers for exceptional efficiency of their stations, life saving, or other acts deserving special mention. To this end keepers will promptly report in detail any act of heroism or gallantry on the part of the personnel of their stations.

When a keeper is commended for the general efficiency of his station duplicate letters shall be sent to each assistant unless reasons exist for contrary action.

Letters of commendation shall remain the personal property of the light keeper concerned, the receipt being noted in the station journal and a copy placed in the records of the inspector's office.

To promote friendly competition in the maintenance of light stations with especial reference to the general efficiency and neatness of the station and personnel, a quarterly circular letter will be issued

Salutes.—Vessels displaying the lighthouse pennant having or board officials or persons entitled to such courtesy, shall be salut with three strokes of the bell, upon arrival at, departure from, o when passing a light station equipped with a bell. Bells havi no inside clapper should have connecting rods detached before salu ing with the striker.

Boats.—Boats shall be marked and known by their official number assigned by the inspector. Boats shall be used for official purpos only, which shall be construed to include all work connected direct or indirectly with the care and maintenance of the station and t personnel thereof.

Station boats must never be used for heavy freight or by worki parties, except under the supervision of the keeper.

Keyboard.—When practicable, a keyboard will be kept in a saf and accessible place where all official keys must be hung by ligh keepers when absent from station.

The official allowance list will be strictly adhered to and no ot supplies, implements, or equipment than those authorized officia l shall be used in the care and maintenance of a light station.

Necessary supplies will be issued daily by the keeper in the pr ence of an assistant, if practicable. The exact quantities thus pended must be entered immediately in the record for entering penditures.

The 5-gallon brass cans shall be used for the daily issue of mine oil and must be completely emptied and cleaned before refilli In refilling, decant all except about 1 inch of the oil from the tra porting can into the brass can. The remaining inch of oil in t transporting can is to be then drawn off in a receptacle kept for purpose and set aside for use in the hand and table lamps.

In drawing oil from the brass can for the daily consumption the lens lamp, again lay aside the last inch of oil in the bottom f this can for use in the hand or table lamps. With oil vapor lan the additional precaution must be taken of passing the oil thro the filter the day before and setting aside the last layer of oil in t bottom of the filter for use in the hand or table lamps.

Oil must never be siphoned from the transporting cans or other method of emptying employed than decanting. The oil m be strained when decanted and again when poured into the lam

Precautions against fire.—Every precaution must be taken aga fire. Fire buckets, when provided, must be kept filled with wat and other apparatus accessible, and in a fixed place.

Liquid chemical fire extinguishers must be kept charged, rechar at least once every year, in a fixed and easily accessible place, and t personnel instructed in their use.

Ashes or sand must be kept available for use on burning min oil, gasoline, etc.

Should oil or gasoline become ignited, the best means of extinguishing it is by the use of the liquid chemical extinguishers, being careful to direct the stream at the base of the fire. In the absence of a proper extinguisher sand, earth, or ashes should be used. A heavy woolen blanket thrown over the fire will also tend to smother it.

[illegible] or lights [illegible] must never be left unattended.

Smoking is prohibited in all buildings belonging to the Lighthouse Service except in living quarters.

Open lights will never be taken into rooms used for storing oils, [illegible] or other inflammable materials.

Only safety matches or those furnished by the Lighthouse Service, will be allowed on lighthouse premises.

Oily waste, rags, and other means of spontaneous combustion will be promptly destroyed.

Uniforms.—Light keepers shall wear the uniform prescribed by the Uniform Regulations and in the manner indicated in Part I, General Instructions to all Employees, page 149.

Rep [illegible] working clothes shall be worn when engaged in dirty work.

Rep [illegible] shall be worn when cleaning lamps and lenses.

Subsistence.—Each appointed light keeper is by law entitled to receive one ration per day, or in the discretion of the Commissioner, commutation [illegible] at the rate of 30 cents per ration. Such provisions [illegible] contracted for by the inspector may be furnished in kind [illegible] where, in the opinion of the inspector, it is for the best interests of the service. Where rations in kind are delivered, the cost [illegible] will be deducted from the allowance of 30 cents per day and the difference paid in cash as a commutation. Where no rations in kind [illegible] issued, the allowance of 30 cents per day will be paid in cash.

Efficiency and commendation.—Letters of commendation may be issued by the inspector to light keepers for exceptional efficiency of their stations, life saving, or other acts deserving special mention. To this end keepers will promptly report in detail any act of heroism or gallantry on the part of the personnel of their stations.

When a keeper is commended for the general efficiency of his station duplicate letters shall be sent to each assistant unless reasons exist for contrary action.

Letters of commendation shall remain the personal property of the light keeper concerned, the receipt being noted in the station journal and a copy placed in the records of [illegible] inspector's office.

To promote friendly competition in the [illegible] tenance of light [illegible] tions with especial reference to the [illegible] iciency and of the station and personnel, a quart[illegible] r letter [illegible]

by the inspector giving names of all light keepers who have been commended during the last quarter.

Efficiency stars.—Light keepers who have been commended for efficiency at each quarterly inspection for a fiscal year shall be entitled to wear the inspector's efficiency star for the succeeding fiscal year.

Light keepers who have been authorized to wear the inspector's efficiency star for three successive years shall be entitled to wear in lieu thereof and for the following fiscal year the Commissioner's efficiency star.

The inspector's efficiency star shall be of gilt and the Commissioner's efficiency star of silver. Stars shall be worn in the manner prescribed in the Uniform Regulations.

Efficiency pennant.—The light station in each district receiving the highest mark for general efficiency during the fiscal year shall be entitled to fly the "efficiency pennant" during the succeeding fiscal year. A circular letter will be issued by the inspector as soon as practicable after July 1, announcing the station winning the efficiency pennant and giving the names of the light keepers attached to that station.

The efficiency pennant shall be similar in size and design to the inspector's pennant and shall never be displayed above or on the same staff as the national colors.

Light keepers shall take precedence in their respective grades and when practicable in the selection for duty, promotion, etc., as follows:

First. Those wearing the Commissioner's efficiency star arranged in order of length of service under that star.

Second. Those wearing the inspector's efficiency star arranged in order of length of service under that star.

Third. Arranged in order of efficiency from official records.

Fourth. Length of service.

No women or children will be allowed to reside at isolated stations, where there are two or more keepers, unless by special permission of the Commissioner previously obtained.

Improper use of dwelling and structures.—No light keeper's dwelling or lighthouse structure shall be used as a pilot station nor as a boarding or lodging place for pilots, or other persons not in the Lighthouse Service, except by special authority of the Commissioner.

Painting.—The following colors shall be used in painting lighthouse structures where no other color is established by proper authority:

Outside colors:

(*a*) Wooden structures.................... Dark red, metallic brown or white, with red or lead colored trimmings.

(*b*) Towers.................................. White.

Outside colors—Continued.

(c) Lanterns and gallery rails	Black.
(d) Iron structures (other than towers)	Brown metallic.
(e) Shutters	Red, green or brown.
(f) Iron walks, rails and steps	Brown metallic.
(g) Wooden walks, rails and steps	Lead color.
(h) Stone and brick work (when authorized) and rough board work	Whitewash.

Inside colors:

(a) Interior of lanterns, and generally for all interior woodwork except hardwood (with the exception of blank panes which should be dull black)	White.
(b) Floors, staircases, and walls, when authorized to be painted. (Hard pine floors and hardwoods generally are not to be painted.)	Grained or lead color.
(c) Iron staircase and railings, and interior ironwork in general, pedestals and service tables	Brown metallic.
(d) Walls, cellars, and outhouses, when painting has not been authorized	Whitewash.
(e) Plastered walls	Light sea green, drab or cream colored transparent oil paint.

Interior walls and ceilings.—Where practicable the interior walls and ceilings of houses shall be painted regulation tints instead of being papered.

Whitewash must never be used on ironwork.

Repairs, etc., by keepers.—All painting and application of washes to structures at light stations, and minor repairs and improvements required in the ordinary preservation and maintenance of the buildings and station shall be done, so far as practicable, by the keepers of the station under the direction of the inspector; but in the case of extensive repairs, or when from any reasonable cause it is impracticable for the keepers to do the work required, it shall be done under the direction of the inspector by competent workmen.

To purify rain water contaminated with chloride of lead from salt spray resting in the leads of structures put a small quantity of pulverized chalk or whiting into the cistern and stir well after each rain.

Gutters and eaves troughs on buildings must be frequently examined, cleaned, and repainted when necessary.

Elevated iron walks must be frequently overhauled to prevent rust around boltheads.

Clean lens and lantern daily.—To clean lens, wipe with soft linen cloth and finally polish with a thoroughly dry buff skin. Remove oil or grease with a linen cloth moistened with diluted ammonia or other authorized wash. Never use a skin which has been wet or damp.

To prevent frosting of lantern glass rub small quantity of glycerin over surface with linen cloth, repeating as necessary.

To clean reflectors dust with feather brush, rub with buff skin lightly dusted with authorized powder kept in a double muslin bag, then rub lightly with a second buff skin and finally with a third by passing over reflector with a light quick circular stroke. Stove gas will tarnish reflectors.

Soiled chimneys should be rubbed with a rag or soft wood dipped in oil. If still discolored, rub with a wet cloth and soda or common salt, afterwards cleanse in warm water as adhering salt will cause breakage.

Dampened cloths and hand brushes should be used in general cleaning about lantern to prevent dust.

Storing cleaning utensils.—The lantern must never be used to store cleaning utensils, etc.

Broken lantern glass must be promptly replaced. After cutting, the edges of a glass should be ground level and smooth by rubbing on a cast-iron plate covered with sharp wet sand or with a block of coarse carborundum. To avoid breakage from oscillation about one-sixteenth of an inch must be left on all sides between the glass and its frame. The glass must be rested on thin sheets of lead or soft wood. In joining two pieces of glass which rest one upon the other, cover the upper edge of the lower piece with putty two-tenths of an inch thick, on this place two small strips of lead and then the upper plate of glass; the weight will press out any excess of putty which should be immediately removed with a glazing knife. Lay the putty outside the frame evenly and flush with the sash. Lock all screw heads by covering with putty.

Revolving clockwork must be kept clean, and well oiled after removing old and gummy oil. Clock oil only should be used on the works, a cloth greased with clock oil on the iron or steel parts. The use of a salted grease is forbidden. The foot of the fly shaft must be watched to prevent cutting or wearing. The weight when not in use is to be kept on its rest.

The chariot or carriage of a revolving lens must be kept clean and the rollers or balls well oiled. When necessary to remove the rollers or balls, do so one at a time; in replacing, put back on same shaft or pin from which removed without changing the number of washers.

Machinery and apparatus.—Instructions for the operation of all machinery and apparatus will be furnished to each station to cover the individual equipment of that station.

Standard time.—In all records at light stations standard time only shall be used, except in off-lying regions where such time system is not in use.

Books.—The following books must be kept in ink, up to date, and as a part of the official records of the station:

Journal.—This shall be a complete record of the important events at the stations having resident keepers, including weather, work in progress, officials visiting the station, visibility of adjacent lights and gas buoys, absences of light keepers, and any irregularity in working of equipment or appliances at the station or noticed in that of other stations. Use as much space on a page as is necessary to record the important events and occurrences for each day.

Expenditure book.—A record of expenditures will be kept at all stations having resident keepers. This may be kept in rough in a copy of Form 30 or in a blank memorandum book, as may be preferred.

Watch book.—This book shall be kept at stations having assistant keepers. In this record shall be entered the working condition of the light and fog signal at the station and the time of starting and stopping each of them. The time of going on and leaving watch must be entered in this book, and the book must be signed by the person standing the watch immediately upon coming off watch.

Fog-signal record book.—This book shall be kept at all stations having fog signals. It shall be kept in accordance with the printed instructions in front of book.

No erasure, alteration or addition will be made in any official record, except by the keeper, who must initial same, adding an explanatory note, if necessary.

No entries or signatures shall be permitted in any of the official record books of the station, except by light keepers or officials of the Lighthouse Service.

Reports and returns.—Keepers shall submit the following reports and returns:

(*a*) Monthly:
- Report of condition of station.
- Report of fog signal.

(*b*) Quarterly:
- Report of principal supplies at station.

(*c*) Annual:
- Property return.
- Requisition for supplies.
- Receipt for annual supplies.
- Survey of public property.

(*d*) When occurring:
- Report of damage or loss of Government property.
- Receipts for special supplies, materials, equipment, etc.
- Receipts for transfer of property.

(*d*) When occurring—Continued.

Life saving or aid extended to persons in distress.
Misconduct or inefficiency of subordinates.
Irregularity in working of aids to navigation.
Unusual occurrences.

Monthly or special reports on the condition of station, when recommending repairs or alterations, shall be explicit as to measurements, estimates of material, cost, and all other necessary details.

Special reports of damage to lens or signal, machinery, or appurtenances, shall give cause, parts damaged, and detailed information to expedite renewal of same.

APPENDIX D

Finishes

Appendix D: Finishes

C.1 Introduction

All the finishes found in the study of the Three Sisters were paint. Study of the paint layers on the Three Sisters was conducted in two segments--an initial examination of paint layers in situ, followed by selective removal of paint samples for further analysis under a 10 to 30 power zoom microscope. Only samples from the exterior were removed for further analysis. Interior paint layering was only examined in situ and the findings are presented in Chapter 3 - Architectural Description of Existing Conditions. This level of investigation is appropriate for the recommended preservation treatment of interiors. If it is determined that the interior of the Beacon will be restored, additional research should be conducted in the construction documents phase.

The initial field examination focused on comparison of paint layers from similar elements of the three towers to differentiate between original and later materials and to identify similar early layers (associated with Lighthouse Service maintenance practice)* compared to different treatments under different owners. Further analysis was conducted to confirm the field findings and to match the colors recommended for restoration treatment to Munsell standards.

C.2 Paint Data

The data gathered through this process are presented in tables at the end of this section. Table 1 shows all the samples which were examined under the microscope. Tables 2 through 6 show all the samples

---

*See Regulations for the United States Lighthouse Service, 1911, pp. 164-5 reproduced as Appendix B.

from a particular surface together. Table 2 shows shingle paint layering. Table 3 is window frames. Table 4 is sash. Table 5 is door frames. Table 6 is doors. The only samples in Table 1 which are not from one of the surfaces covered by the other tables are P003A and P003B. Both are from the canvas covering of the parapet on the Beacon. The canvas itself has green vertical stripes on top of the natural canvas color. The stripes are about 1½ inches wide with ½ inch of natural canvas exposed between stripes. The canvas does not appear to have been a waterproof finish material prior to painting because the canvas is not saturated wth dried oil as it would be had it been prepared like a floorcloth for water resistance. The paint layers applied to the canvas are shown in Table 1. The discrepancies between the two samples are probably due to irregular chipping and peeling of old layers prior to repainting. All of the colors listed were probably finishes at some time, but the sequence of the more recent layers is difficult to determine. Fortunately our historic period is early in the life of the structure and the historic paint would have been either the black bituminous layer or the dark green. Since the only other colors used on the lighthouses in the historic period are shades of gray, and dark green appears on other elements of the Beacon after the historic period, and the Regulations for the United States Lighthouse Service state that the exteriors of lanterns are to be painted black, we recommend restoration to the black bituminous layer.

Table 2 shows the paint layering found on all the shingles that were examined. In every case the first layer was a light color--white, light gray, light yellow gray, or light yellow. The differences are probably due to different exposure, weathering, soiling and fading. Prior to these effects, the paint was probably white or very light gray. The many layers of the same or very similar colors reflect the maintenance practices of the Lighthouse Service. The later whites, dark grays and colors are probably the work of the private owners. The following observations support this hypothesis

1. There is only minimal soiling between the many early layers of similar color.
2. The early layers have retained their integrity (evidenced by lack of cracking and intrusion of later paints into early layers).

3. Later layers are quite soiled and there are many cracks and intrusions of later paint, indicating long exposure to weather.
4. The Lighthouse Service had standard regulations for paint colors to be used on lighthouses.

Twelve to fourteen layers were applied to the Beacon by the Lighthouse Service in their thirty year tenure compared to three or four by private owners in a sixty year period. The Munsell designation of the recommended paint color for restoration is N 8.5/.

Table C3 shows the paint layering found on the window frames of the three towers. The distinction between Lighthouse Service paint colors and private owners' paints is very clear in these samples. Sample P004C has thirteen layers of gray paint followed by a light blue-green and then white. The final two layers are an incomplete representation of the later paint schemes which are present in sample P002--dark green, medium red orange, medium green, light blue-green, and bright green. A gray paint, Munsell designation N4.5/ is recommended for restoration.

Table 4 shows the paint layering found on the sash. The putty from the original sash in the South Twin (P005B) has the most complete layering. The putty itself has an integral medium red orange (rust) color followed by two layers of light yellow gray, dark green, white, light blue, white, and glossy black. The rust color also occurs as a thin layer, like a stain, on the wood. The light yellow gray matches the shingle color and occurs at the right place in the paint sequence to correspond to the historic period. As with the shingles, the yellowing is probably the result of deterioration of the linseed oil medium, so the recommended paint color will be a light gray. The Munsell designation for the recommended restoration paint is N8.5/.

Table C5 shows the paint layering found on the door frames of the three towers. The medium gray layer, Munsell designation N6.5/, is recommended for restoration.

Table C6 shows the paint layering on the doors of the Twins. In both cases the layering matches the surrounding frame. The medium gray layer, Munsell designation N6.5/, is recommended for restoration.

Table C7 shows the paint layering on the cornices of the three lighthouses. The gray layer, Munsell designation N4.5/, is recommended for restoration.

C.3 Summary of Recommended Finishes

The following chart summarizes the recommended restoration paint colors for the Three Sisters Lighthouses and companies them with the colors specified in the 1911 Regulations . . .

SURFACE	COLOR	MUNSELL DESIGNATION	1911 REGULATIONS
Shingles	Light gray	N8.5/	dark red, metallic brown or white
Window frames	Gray	N4.5/	red or lead
Sash	Light gray	N8.5/	na
Door frames	Medium gray	N6.5/	red or lead
Doors	Medium gray	N6.5/	red or lead
Cornice	Gray	N4.5/	red or lead
Parapet	Black Bituminous	N2/	black

TABLE C1

	P001A	P001B	P001C	P002	P003A	P003B	P004A	P004B	P004C	P005A
SUBSTRATE	WD. SHGL.	WD.SHNGL.	WD.SHNGL.	WOOD	CANVAS	CANVAS	WOOD	WOOD	WOOD	PUTTY
LAYER 1	LT.YL.GR.	LT.GR.	WHITE	MD.GR.	*BLK.BIT.	*BLK.BIT.	WHITE	GRAY	GRAY	MD.RD.OR.
LAYER 2	LT.YL.GR.	MD.GR.	WHITE	LT.GR.	DK.GRN.	DK.GRN.	MD.GR.	GRAY	GRAY	LT.YL.GR.
LAYER 3	LT.YL.GR.	LT GR.	WHITE	DK GR.	LT GR.	YL.	LT.GR.	GRAY	GRAY	LT.YL.GR.
LAYER 4	LT.YL.GR.	WHITE	WHITE	MD.RD.GR.	WHITE	ALUM.	LT.BL.GRN.	GRAY	GRAY	DK.GRN.
LAYER 5	LT.YL.GR.	GREEN	LT.GR.	MD.GRN.	V.DK.GRN.		WHITE	GRAY	GRAY	
LAYER 6	LT.YL.GR.	MD RD GR.		LT.BL.GRN.				GRAY	*GRAY	GL. BLK.
LAYER 7	LT.YL.GR.	WHITE		BRT.GRN.				GRAY	GRAY	WHITE
LAYER 8	*LT.YL.GR.	GREEN						LT. BL. GRN.	GRAY	GL.BLK.
LAYER 9	MD.YL.GR.	WHITE						WHITE	GRAY	
LAYER 10	LT.YL.GR.	GREEN							GRAY	
LAYER 11	MD.YL.GR.	WHITE							GRAY	
LAYER 12	MD.YL.GR.								GRAY	
LAYER 13	LT.GR.								GRAY	
LAYER 14	MD.GR.								LT.B.GRN.	
LAYER 15	WHITE								WHITE	
LAYER 16	WHITE									
LAYER 17	WHITE									

ABBREVIATIONS:

ALUM.	=	ALUMINUM
BIT.	=	BITUMINOUS
BL.	=	BLUE
BLK.	=	BLACK
BRT.	=	BRIGHT
DK.	=	DARK
GL.	=	GLOSSY
GR.	=	GRAY
GRN.	=	GREEN
LT.	=	LIGHT
MD.	=	MEDIUM
OL.	=	OLIVE
OR.	=	ORANGE
RD.	=	RED
SHGL.	=	SHINGLE
V.	=	VERY
WD.	=	WOOD
YL.	=	YELLOW

*denotes layer recommended for restoration

TABLE C1 (cont.)

P005B	P006	P007	P008	P009	P010	P011	P012	P013	P014	P015
PUTTY	PUTTY	PUTTY	WD.SHNGL.	WOOD	WOOD	WOOD	WOOD	WOOD	WOOD	WOOD
MD.RD.OR.	WHITE	GRAY	WHITE	LT.YL.GR.	MD.GR.	LT.GR.	LT.YL.GR.	MD.GR.	LT.GR.	WHITE
LT.YL.GR.	BLACK	OL.GRN.	LT.GR.	WHITE	LT.YL.GR.	WHITE	LT.GR.	WHITE	LT.YL.GR.	BLK.
LT.YL.GR.		GRAY	MD.GR.	WHITE	WHITE	LT.YL.GR.	LT. PINK GR.	LT.YL.GR.	WHITE	LT.GR.
DK.GRN.		YELLOW	WHITE	(INC.)	WHITE	LT. PINK GR.		WHITE	WHITE	DK.GR.
WHITE		DK.GRN.	WHITE		WHITE	WHITE			LT.GR.	BLK.
LT.BL.		GRAY	WHITE		WHITE				WHITE	GRN.
WHITE			WHITE		WHITE				WHITE	BLK.
GL.BLK.			DK.GRN.		WHITE					GRN.
			WHITE		WHITE					WHITE

TABLE C2

SUMMARY OF PAINT
STUDY BY SURFACE

SURFACE STRUCTURE	SHINGLES BEACON	SHINGLES N. TWIN	SHINGLES S. TWIN	SHINGLES BEACON	SHINGLES BEACON	SHINGLES BEACON	SHINGLES BEACON
SAMPLE #	NONE	NONE	NONE	P001A	P001B	P001C	P008
LAYERS:	LT.YL.GR.	LT.YL.	WHITE	*LT.YL.GR.	LT.GR.	WHITE	WHITE
	LT.YL.GR.	LT.YL.	WHITE	LT.YL.GR.	MD.GR.	WHITE	LT.GR.
	LT.YL.GR.	LT.YL.	WHITE	LT.YL.GR.	LT.GR.	WHITE	MD.GR.
	LT.YL.GR.	LT.YL.	WHITE	LT.YL.GR.	WHITE	WHITE	WHITE
	LT.GR.	LT.GR.	MANY LAYERS	LT.YL.GR.	GREEN	LT.GR.	WHITE
	LT.GR.	LT.GR.		LT.YL.GR.	MD.RD.OR.	incomplete	WHITE
	LT.GR.	LT.GR.		LT.YL.GR.	WHITE		WHITE
	WHITE	WHITE		LT.YL.GR.	GREEN		DK.GRN.
	DK.GR.			MD.YL.GR.	WHITE		WHITE
	WHITE			LT.YL.GR.	GREEN		
				MD.YL.GR.	WHITE		
				MD.YL.GR.	TRIM colors		
				LT.GR.	mixed in		
				MD.GR.			
				WHITE			
				WHITE			
				WHITE			

*denotes layer recommended for restoration

TABLE C3

SUMMARY OF PAINT

STUDY BY SURFACE

SURFACE	WINDOW FRAME	WINDOW FRAME	WINDOW FRAME	WINDOW FRAME	WINDOW FRAME	WINDOW FRAME	WINDOW FRAME	WINDOW FRAME
STRUCTURE	BEACON	BEACON	BEACON	BEACON	BEACON	BEACON	NORTH TWIN	NORTH TWIN
DATE?	ORIGINAL	ORIGINAL	ORIGINAL	ORIGINAL	ORIGINAL	LATER	ORIGINAL	LATER
SAMPLE #	NONE	P002	P004A	P004B	P004C	NONE	NONE	NONE
LAYER 1	MD.GR.	MD.GR.	WHITE	GRAY	*GRAY	DK.BRN.	MD.GR.	WHITE
LAYER 2	DK.GRN.	LT.GR.	MD GR.	GRAY	GRAY	MD.RD.OR.	MD.GR.	GRN.
LAYER 3	MD.RD.OR.	DK.GRN.	LT.GR.	GRAY	GRAY	BRT.GRN.	MD.GR.	WHITE
LAYER 4	BRT.GRN.	MD.RD.OR.	LT.BL.GRN.	GRAY	GRAY		#?	
LAYER 5	Incomplete	MD.GRN.	WHITE	GRAY	GRAY		WHITE	
LAYER 6		LT.BL.GRN.	Incomplete	GRAY	GRAY		GRN.	
LAYER 7		BRT. GRN.	layering	GRAY	GRAY		WHITE	
LAYER 8				LT.BL.GRN.	GRAY			
LAYER 9				WHITE	GRAY			
LAYER 10					GRAY			
LAYER 11					GRAY			
LAYER 12					GRAY			
LAYER 13					GRAY			
LAYER 14					LT.BL.GRN.			
LAYER 15					WHITE			

*denotes layer recommended for restoration

TABLE C4

SUMMARY OF PAINT

STUDY BY SURFACE

SURFACE	SASH	ASH	SASH	ASH	SASH	ASH	SASH PUTTY	SASH PUTTY	ASH PUT
STRUCTURE	BEACON	BEACON	N. TWIN	N. TWIN	S. TWIN	S. TWIN	S. TWIN	S. TWIN	S. TWIN
DATE?	ORIGINAL	LATER	ORIGINAL	LATER	ORIGINAL	LATER	ORIGNIAL	ORIGINAL	LATER
SAMPLE #	NE	NE	NONE	NE	NONE	ONE	P005A	P005B	P006
LAYER 1	MD.GR.	WHITE	RED	WHITE	RED	WHITE	MD.RD.OR.	MD.RD.OR.	WHITE
LAYER 2	WHITE		WHITE	GN	WHITE	GN	LT.YL.GR.	LT.YL.GR.	BLACK
LAYER 3			WHITE	WHITE?	WHITE	WHITE	*LT.YL.GR.	LT.YL.GR.	
LAYER 4			WHITE	BLACK	WHITE	BK	DK.GRN.	DK.GRN.	
LAYER 5			#?		#?		WHITE	WHITE	
LAYER 6			GREEN		GREEN		GL.BLK.	LT.BL.	
LAYER 7			BLACK		BLACK			WHITE	
LAYER 8								GL.BLK.	

*denotes layer recommended for restoration

TABLE C5

SUMMARY OF PAINT
STUDY BY SURFACE

SURFACE	DOOR FRAME	DOOR FRAME	DOOR FRAME
STRUCTURE	BEACON	N. TWIN	S. TWIN
DATE?	ORIGINAL	ORIGINAL	ORIGINAL
SAMPLE #	P009	NONE	NONE
LAYER 1	*DK.GR.	GRAY	MD.GR.
LAYER 2	MD.GR.	LT.BRN.	WHITE
LAYER 3	CREAM	BROWN	WHITE
LAYER 4	WHITE	CREAM	
LAYER 5	WHITE	WHITES	

*denotes layer recommended for restoration

TABLE C6

SUMMARY OF PAINT
STUDY BY SURFACE

SURFACE	DOOR	DOOR
STRUCTURE	N. TWIN	S. TWIN
DATE?	ORIGINAL	ORIGINAL
SAMPLE #	NONE	NONE
LAYER 1	GRAY	MD.GR.
LAYER 2	LT.BRN.	LT.BL.
LAYER 3	BROWN	WHITE
LAYER 4	MD.BL.	
LAYER 5	WHITE	

APPENDIX E

Hardware

Appendix E: Hardware

The original hardware of the Three Sisters has survived to the extent that a representative sample of most elements is still available to be documented and the majority of the individual pieces are still in place and functional.

D.1 Door Hardware

The main entrance doors to the Three Sisters swing on 4 by 4 inch cast iron butt hinges, two per door. Cast iron mortise locks with ceramic knobs, 2 inch diameter cast iron roses, rectangular keyhole escutcheons, latch, and deadbolt were used to secure the doors.

The hatch door in the floor of the lantern was hung on two small butt hinges mounted on the face of the door. A modern hasp was used to secure the hatch from below. The interior door to the lantern deck hung on 2 by 2 inch surface-mounted butt hinges and was secured with a small flush-mortised spring latch. The exterior door to the lantern deck hangs on two T-strap surface-mounted hinges. It is secured with a plate-mounted hook and eye.

D.2 Window Hardware

The historic windows are six-over-six single-hung sash. Sash fasteners are the helical spring type similar to Judd's patent as illustrated on page 77 in the Russell and Erwin Catalogue of American Hardware (see Figure 16). The sash pulleys are the standard iron sheaves mortised into the jambs. Historic pulleys have rectangular faceplates; the nonhistoric faceplates have rounded ends. Shutter hardware was found on the South Twin, but it is considered nonhistoric because there is no evidence of shutter hardware on the other towers and the historic photographs do not show shutters.

Figure 16

Historic sash lock (two views) in the North Twin
Photograph by W. Howell June 1983

APPENDIX F

Recommendations for Tie-down of Beacon (memo 2/3/84)

FEB 3 1984

H30 (DSC-TNE)

Memorandum

To: Superintendent, Cape Cod National Seashore

From: Assistant Manager, Northeast Team, Denver Service Center

Reference: Cape Cod, Pkg. No. 138, Three Sisters Lighthouses, Historic Structure Report

Subject: Design for Soil Anchors and Connectors

In consultation with Denver Service Center Structural Engineers Terry Wong and Maurice Paul, Project Historical Architect William Howell has prepared the following recommendations for anchoring the Beacon to prevent overturning.

In order to resist wind loading on the Beacon in its temporary location, we recommend four equally spaced soil anchors (Chance Catalog Number 1(146 or equal) connected to the sill as shown on the enclosed drawing. (A. B. Chance Company, 210 North Allen Street, Centralia, Missouri 65240).

This recommendation is based on 100 MPH winds and the present configuration of the cribbing. Anchors should be centered between cribbing. There are innumerable possibilities for the design of connections between the sill and the soil anchors. For ease of installation, we suggest a steel "T" lag-bolted to the bottom of the sill as shown on the enclosed detail and connected to the eye of the soil anchor by a long turnbuckle such as 3001T56 found on Page 189C of the 1982 McMaster-Carr catalog and priced at $7.29 each. The steel "Ts" should be WT5x11 sections, 6 inches long with four bolt holes at 3 inches on center in the flange and one hole in the stem large enough to accommodate the pin of the turnbuckle and at least 1-1/8 inches from the edge of the "T". Lag bolts should be 5/8-inch diameter and 5 inches long. The depth of penetration of the soil anchors may be adjusted so that the turnbuckles fit between the Tripleye on the anchor and the hole in the steel "T".

(sgd) Gerald D. Patten

Gerald D. Patten

Enclosures

cc:
Reg. Dir., North Atlantic, w/o encs.
NARO-PC-Mr. Skelton, w/encs.

bcc:
DSC-TSE-Mr. McGrath, w/encs.
DSC-TNE-PIFS, w/o encs.

TNE:WHowell:hca:2/1/84:5545

NO-WRENCH SCREW ANCHOR

- **For Hand or Machine Installation**

The Chance No-Wrench Screw Anchor may be installed by hand or by machine The Thimbleye* or Tripleye* on the rod has a large opening to admit a turning bar for screwing the anchor down. The eye will also fit into an adapter available from most hole boring machine manufacturers so that the anchor may be power installed. The No-Wrench Screw Anchor consists of a drop forged, steel Thimbleye or Tripleye rod, electric welded to a steel helix. The entire anchor is hot dip galvanized for long resistance to rust

No wrench screw anchors may be installed to a greater depth to reach a more firm soil by using a 6-ft extension rod with forged coupling and a forged tripleye nut, catalog number 402

Catalog numbers 4345, 6346 and 816 may be ordered with a forged thimbleye nut rather than the standard tripleye. To order a thimbleye nut simply add "-1" to the suffix of the catalog number. Example: Catalog No 6346-1.

ORDERING INFORMATION

Catalog No. Tripleye	Anchor Size Dia.	Rod Dia. and Length
4345	4 "	¾" x 54"
6346	6 "	¾" x 66"
816	8 "	1" x 66"
10146	**10 "**	**1¼" x 66"**
10148	10 "	1¼" x 96"
12537	13½"	1¼" x 96"
15148	15"	1¼" x 96"

1¼" x 6' Tripleye Extension — Cat No 402 — 2900 Lbs Per 100 Pieces.

SWAMP SCREW ANCHOR

- **For Difficult Swamp Guying**

Chance Swamp Screw Anchors are specifically designed for installation in swamps, bogs, and marshes They may be installed to the necessary depth to reach firm soil by using pipe wrenches, chain type pipe tongs, or, where circumstances permit, by hole boring equipment

- **Threaded For Standard Pipe**

Anchor hubs and nuts are threaded to receive API threaded pipe to make a secure assembly Extra sections of pipe may be added for deeper penetration, often necessary to obtain sufficient holding capacity.

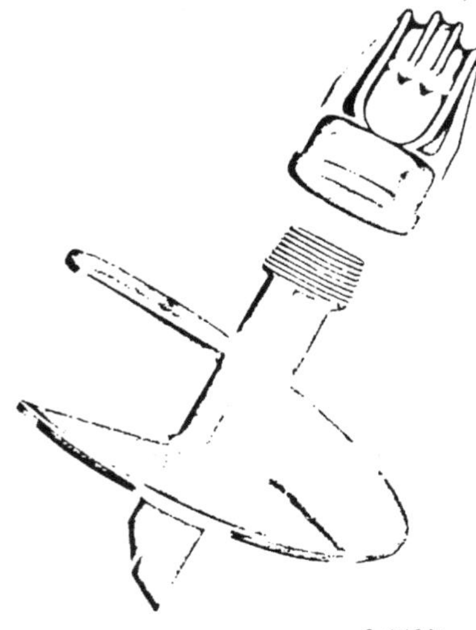

ORDERING INFORMATION

Cat No Thimbleye Nut	Cat. No. Tripleye Nut	Anchor Size	Pipe Size (Not Included)	Approx. Weight Per 100
10150-AS-1	10150-AS	10"	1½"	1300
132-AS-1	132-AS	13½"	2 "	2150
152-AS-1	152-AS	15"	2 "	3150

G 10M 9-81
JANUARY, 1981
Rev. 9-81

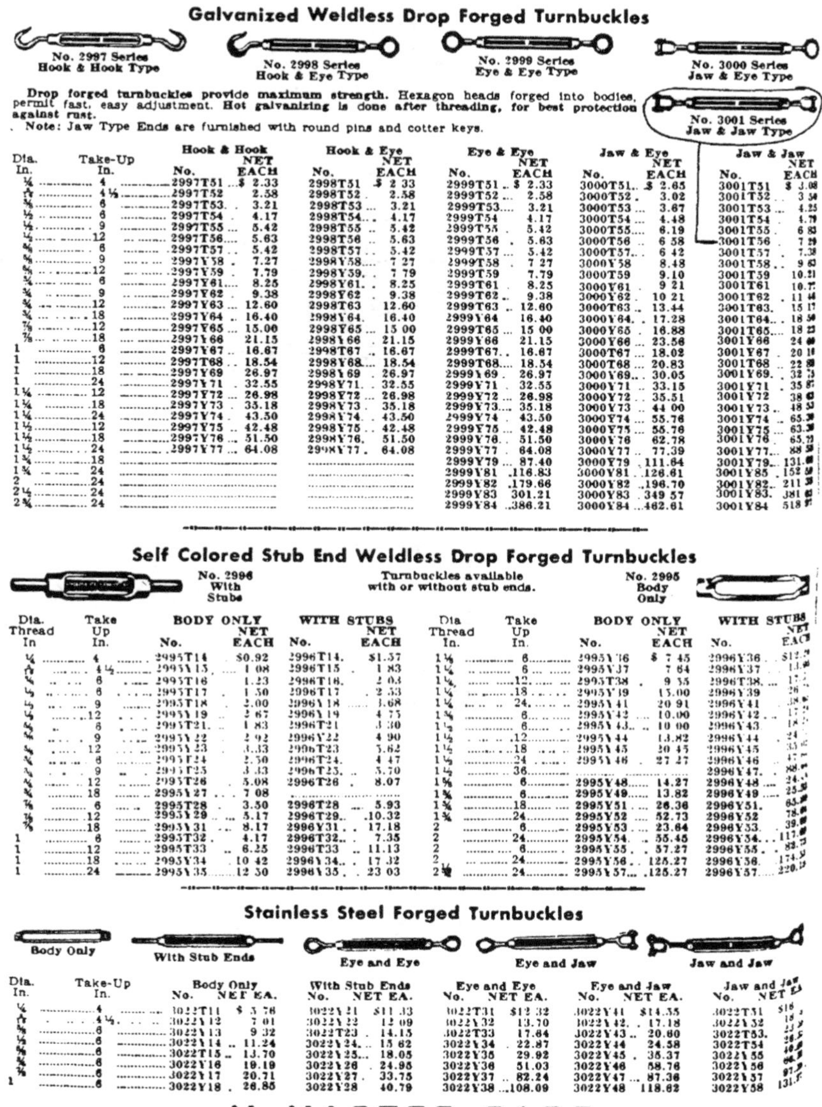

TURNBUCKLES

GALVANIZED—WELDLESS
DROP FORGED—STAINLESS STEEL

Galvanized Weldless Drop Forged Turnbuckles

No. 2997 Series Hook & Hook Type

No. 2998 Series Hook & Eye Type

No. 2999 Series Eye & Eye Type

No. 3000 Series Jaw & Eye Type

No. 3001 Series Jaw & Jaw Type

Drop forged turnbuckles provide maximum strength. Hexagon heads forged into bodies, permit fast, easy adjustment. **Hot galvanizing is done after threading, for best protection against rust.**

Note: Jaw Type Ends are furnished with round pins and cotter keys.

Dia. In.	Take-Up In.	Hook & Hook No.	Hook & Hook NET EACH	Hook & Eye No.	Hook & Eye NET EACH	Eye & Eye No.	Eye & Eye NET EACH	Jaw & Eye No.	Jaw & Eye NET EACH	Jaw & Jaw No.	Jaw & Jaw NET EACH
¼	4	2997T51	$ 2.33	2998T51	$ 2.33	2999T51	$ 2.33	3000T51	$ 2.65	3001T51	$ 3.08
5/16	4½	2997T52	2.58	2998T52	2.58	2999T52	2.58	3000T52	3.02	3001T52	3.54
⅜	6	2997T53	3.21	2998T53	3.21	2999T53	3.21	3000T53	3.67	3001T53	4.25
½	6	2997T54	4.17	2998T54	4.17	2999T54	4.17	3000T54	4.48	3001T54	4.79
½	9	2997T55	5.42	2998T55	5.42	2999T55	5.42	3000T55	6.19	3001T55	6.83
½	12	2997T56	5.63	2998T56	5.63	2999T56	5.63	3000T56	6.58	3001T56	7.29
⅝	6	2997T57	5.42	2998T57	5.42	2999T57	5.42	3000T57	6.42	3001T57	7.38
⅝	9	2997Y58	7.27	2998Y58	7.27	2999T58	7.27	3000Y58	8.48	3001T58	9.63
⅝	12	2997Y59	7.79	2998Y59	7.79	2999T59	7.79	3000T59	9.10	3001T59	10.21
¾	6	2997Y61	8.25	2998Y61	8.25	2999T61	8.25	3000Y61	9.21	3001T61	10.7[illegible]
¾	9	2997Y62	9.38	2998Y62	9.38	2999T62	9.38	3000Y62	10.21	3001T62	11.44
¾	12	2997Y63	12.60	2998T63	12.60	2999T63	12.60	3000T63	13.44	3001T63	15.17
¾	18	2997Y64	16.40	2998Y64	16.40	2999Y64	16.40	3000Y64	17.28	3001T64	18.50
⅞	12	2997Y65	15.00	2998Y65	15.00	2999T65	15.00	3000Y65	16.88	3001T65	18.23
⅞	18	2997Y66	21.15	2998Y66	21.15	2999Y66	21.15	3000Y66	23.56	3001Y66	24.40
1	6	2997Y67	16.67	2998T67	16.67	2999T67	16.67	3000T67	18.02	3001Y67	20.18
1	12	2997T68	18.54	2998Y68	18.54	2999T68	18.54	3000T68	20.83	3001T68	22.88
1	18	2997Y69	26.97	2998Y69	26.97	2999Y69	26.97	3000Y69	30.05	3001Y69	32.73
1	24	2997Y71	32.55	2998Y71	32.55	2999Y71	32.55	3000Y71	33.15	3001Y71	35.87
1¼	12	2997Y72	26.98	2998Y72	26.98	2999Y72	26.98	3000Y72	35.51	3001Y72	38.63
1¼	18	2997Y73	35.18	2998Y73	35.18	2999Y73	35.18	3000Y73	44.00	3001Y73	48.53
1¼	24	2997Y74	43.50	2998Y74	43.50	2999Y74	43.50	3000Y74	55.76	3001Y74	65.[illegible]
1½	12	2997Y75	42.48	2998Y75	42.48	2999Y75	42.48	3000Y75	55.76	3001Y75	63.[illegible]
1½	18	2997Y76	51.50	2998Y76	51.50	2999Y76	51.50	3000Y76	62.78	3001Y76	65.[illegible]
1½	24	2997Y77	64.08	2998Y77	64.08	2999Y77	64.08	3000Y77	77.39	3001Y77	88.[illegible]
1¾	18					2999Y79	87.40	3000Y79	111.64	3001Y79	131.[illegible]
1¾	24					2999Y81	116.83	3000Y81	126.61	3001Y85	152.[illegible]
2	24					2999Y82	179.66	3000Y82	196.70	3001Y82	211.[illegible]
2¼	24					2999Y83	301.21	3000Y83	349.57	3001Y83	381.[illegible]
2¾	24					2999Y84	386.21	3000Y84	462.61	3001Y84	518.[illegible]

Self Colored Stub End Weldless Drop Forged Turnbuckles

No. 2996 With Stubs

Turnbuckles available with or without stub ends.

No. 2995 Body Only

Dia. Thread In	Take Up In.	BODY ONLY No.	BODY ONLY NET EACH	WITH STUBS No.	WITH STUBS NET EACH
¼	4	2995T14	$0.92	2996T14	$1.57
5/16	4½	2995Y15	1.08	2996T15	1.83
⅜	6	2995T16	1.23	2996T16	2.03
½	6	2995T17	1.50	2996T17	2.53
½	9	2995T18	2.00	2996Y18	3.68
½	12	2995Y19	2.67	2996Y19	4.75
⅝	6	2995T21	1.83	2996T21	3.30
⅝	9	2995Y22	2.92	2996Y22	4.90
⅝	12	2995Y23	3.33	2996T23	5.62
¾	6	2995T24	2.50	2996T24	4.47
¾	9	2995T25	3.33	2996T25	5.70
¾	12	2995T26	5.08	2996T26	8.07
¾	18	2995Y27	7.08		
⅞	6	2995T28	3.50	2996T28	5.93
⅞	12	2995Y29	5.17	2996T29	10.32
⅞	18	2995Y31	8.17	2996Y31	17.18
1	6	2995T32	4.17	2996T32	7.35
1	12	2995T33	6.25	2996T33	11.13
1	18	2995Y34	10.42	2996Y34	17.32
1	24	2995Y35	12.50	2996Y35	23.03
1⅛	6	2995Y36	$ 7.45	2996Y36	$12.[illegible]
1¼	6	2995Y37	7.64	2996Y37	13.[illegible]
1¼	12	2995T38	9.55	2996T38	17.[illegible]
1¼	18	2995Y39	15.00	2996Y39	26.[illegible]
1¼	24	2995Y41	20.91	2996Y41	38.[illegible]
1⅜	6	2995Y42	10.00	2996Y42	17.[illegible]
1½	6	2995Y43	10.00	2996Y43	18.[illegible]
1½	12	2995Y44	13.82	2996Y44	24.[illegible]
1½	18	2995Y45	20.45	2996Y45	35.[illegible]
1½	24	2995Y46	27.27	2996Y46	47.[illegible]
1½	36			2996Y47	88.[illegible]
1⅝	6	2995Y48	14.27	2996Y48	24.[illegible]
1¾	6	2995Y49	13.82	2996Y49	25.[illegible]
1¾	18	2995Y51	26.36	2996Y51	65.[illegible]
1¾	24	2995Y52	52.73	2996Y52	78.[illegible]
2	6	2995Y53	23.64	2996Y53	39.[illegible]
2	24	2995Y54	55.45	2996Y54	117.[illegible]
2	6	2995Y55	57.27	2996Y55	82.[illegible]
2	24	2995Y56	125.27	2996Y56	174.[illegible]
2½	24	2995Y57	125.27	2996Y57	220.[illegible]

Stainless Steel Forged Turnbuckles

Body Only — With Stub Ends — Eye and Eye — Eye and Jaw — Jaw and Jaw

Dia. In.	Take-Up In.	Body Only No.	Body Only NET EA.	With Stub Ends No.	With Stub Ends NET EA.	Eye and Eye No.	Eye and Eye NET EA.	Eye and Jaw No.	Eye and Jaw NET EA.	Jaw and Jaw No.	Jaw and Jaw NET EA.
¼	4	3022T11	$ 5.78	3022Y21	$11.33	3022T31	$12.32	3022Y41	$14.55	3022T51	$18.[illegible]
5/16	4½	3022Y12	7.01	3022Y22	12.09	3022Y32	13.70	3022Y42	17.18	3022Y52	19.[illegible]
⅜	6	3022Y13	9.32	3022T23	14.15	3022T33	17.64	3022Y43	20.60	3022T53	23.[illegible]
½	6	3022Y14	11.24	3022Y24	15.62	3022Y34	22.87	3022Y44	24.58	3022T54	28.[illegible]
⅝	6	3022T15	13.70	3022Y25	18.05	3022Y35	29.92	3022Y45	35.37	3022Y55	40.[illegible]
¾	6	3022Y16	19.19	3022Y26	24.95	3022Y36	51.03	3022Y46	58.76	3022Y56	66.[illegible]
⅞	6	3022Y17	20.71	3022Y27	33.75	3022Y37	82.24	3022Y47	87.36	3022Y57	97.[illegible]
1	6	3022Y18	26.85	3022Y28	40.79	3022Y38	108.09	3022Y48	118.62	3022Y58	131.[illegible]

McMASTER-CARR

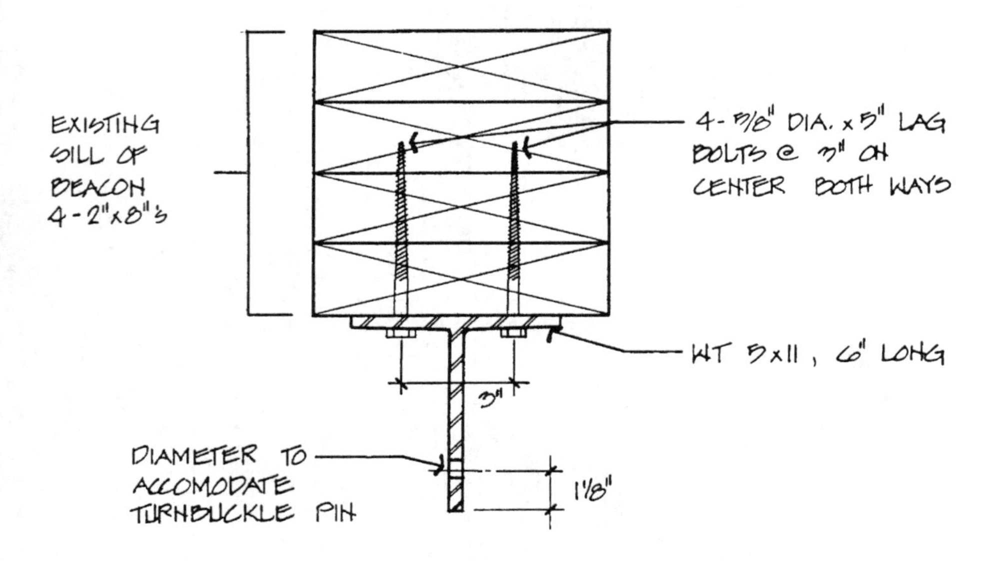
EXISTING
SILL OF
BEACON
4-2"x8"s
4-5/8" DIA. x 5" LAG
BOLTS @ 3" ON
CENTER BOTH WAYS
WT 5x11, 6" LONG
3"
DIAMETER TO
ACCOMODATE
TURNBUCKLE PIN
1 1/8"

APPENDIX G

Annotated Bibliography

APPENDIX G
Annotated Bibliography

The major portion of lighthouse records dealing with the Nauset Beach lights between 1852 and 1921 were burned in the Commerce Department fire of 1921. As a result, Record Group 26, Records of the United States Coast Guard in the National Archives has a restricted number of documents. The Cartographic and Architectural Division of that archives had no drawings on the Nauset Beach lights. Several pictures were obtained from the Still Photo Division. Few records on those lights were found in the United States Coast Guard Academy Library at New London, Connecticut. The library contained several maps of the property showing building locations and elevation and floor plan drawings of the 1875 wood frame keeper's dwelling. The First District Coast Guard office in Boston had a document giving a good description of the lights in 1910. Cape Cod National Seashore files yielded some material, including several maps and photographs. Those depositories investigated, which yielded no primary documents on the Nauset Beach lights, included the Massachusetts Historical Society, Bostonian Society, Boston Public Library, Massachusetts Historical Commission, Massachusetts State Archives, and the Library of Congress.

Many secondary sources gave only general descriptions of the Nauset Beach lights, such as Edward R. Snow's, _Famous New England Lighthouses_ and Malcolm F. Willoughby's, _Lighthouses of New England_. These works proved of little value and were not used for source material. Only those books cited below were used and they contained mostly background information on the Lighthouse Service.

Primary Sources

Unpublished documents

National Archives, Washington, D.C.

Clipping File. Records of the United States Coast Guard. Record Group 26. This file has some information on the Nauset Beach lights between 1860 and 1900 in a folder marked "Nauset Beach Lights." Included are appropriate sections from the Lighthouse Board Annual Reports.

Correspondence of the Bureau of Lighthouses, 1911-39. Records of the United States Coast Guard. Record Group 26. This series of documents contains only a small amount of material on the Nauset Beach lights.

Description of Light Stations 1876-1938. Records of the United States Coast Guard. Record Group 26. This entry contained a 1927 description of the Nauset Beach lights.

Index to Lighthouse Board Letterbooks (Nauset Beach lights). Records of the United States Coast Guard. Record Group 26. All that remains from the Commerce Department fire of 1921 are the index to these letterbooks. The one or two index sentences describing each letter's contents provide some information, but do not give a full account of each letter's subject matter.

Letter from the Secretary of the Treasury transmitting the Report of the General Superintendent of the Lighthouse Establishment, December 19, 1850. Records of the United States Coast Guard. Record Group 26. The report contains a list of supplies delivered to Nauset Beach in 1850.

Lighthouse Superintendent, Boston 1826-44 (correspondence). Records of the United States Coast Guard. Record Group 26. This group of documents has a small amount of material pertaining to the early Nauset Beach lights.

United States Coast Guard Academy, New London, Connecticut

Group Plan Map of Nauset Beach structures. No date.

First District Coast Guard, Boston, Massachusetts

Description of Lighthouse Towers, Buildings, and Premises at Nauset Beach, Massachusetts, Light-Station. January 11, 1910. The document gives a fairly good description of the three lights as to their dimensions including doors and windows.

Cape Cod National Seashore

History of "Jack O'Lantern" -- Hall Cottages. A telephone conversation between Doris Doane (a Cape Cod National Seashore employee) and Harold Hall. November 21, 1975. Hall tells of this parents' purchase of the third of the three sisters in 1923.

John A. Cummings's written account of the sale of two of the "three sisters" to his parents. March 23, 1969. Cummings gave an account of when his parents bought the lights, the cost, and what they did with them.

Published Government Documents

Annual Report of the Light-House Board to the Secretary of the Treasury for the Year 1868. Washington, D.C.: Government Printing Office, 1868.

Annual Report of the Light-House Board to the Secretary of the Treasury for the Fiscal Year 1869. Washington, D.C.: Government Printing Office, 1869.

Annual Report of the Light-House Board of the United States, 1873. Washington, D.C.: Government Printing Office, 1873.

Annual Report of the Light-House Board of the United States to the Secretary of the Treasury for the Fiscal Year Ending June 30, 1874. Washington, D.C.: Government Printing Office, 1874.

Annual Report of the Light-House Board to the Secretary of the Treasury for the Fiscal Year Ending June 30, 1876. Washington, D.C.: Government Printing Office, 1877.

Annual Report of the Light-House Board to the Secretary of the Treasury for the Fiscal Year Ending June 30, 1882. Washington, D.C.: Government Printing Office, 1882.

Annual Report of the Light-House Board to the Secretary of the Treasury for the Fiscal Year Ending June 30, 1892. Washington, D.C.: Government Printing Office, 1892.

Annual Report of the Light-House Board to the Secretary of the Treasury for the Fiscal Year Ending June 30, 1895. Washington, D.C.: Government Printing Office, 1895.

Annual Report of the Commissioner of Lighthouses to the Secretary of Commerce and Labor for the Fiscal Year Ending June 30, 1911. Washington, D. C.: Government Printing Office, 1911.

Annual Report of the Commissioner of Lighthouses to the Secretary of Commerce and Labor for the Fiscal Year Ending June 30, 1912. Washington, D.C.: Government Printing Office, 1913.

Annual Report of the Commissioner of Lighthouses to the Secretary of Commerce for the Fiscal Year Ending June 30, 1923. Washington,

Primary Sources

Unpublished documents

National Archives, Washington, D.C.

Clipping File. Records of the United States Coast Guard Record Group 26. This file has some information on the Nauset Beach lights between 1860 and 1900 in a folder marked "Nauset Beach Lights." Included are appropriate sections from the Lighthouse Board Annual Reports.

Correspondence of the Bureau of Lighthouses, 1911-39. Records of the United States Coast Guard. Record Group 26. This series of documents contains only a small amount of material on the Nauset Beach lights.

Description of Light Stations 1876-1938. Records of the United States Coast Guard. Record Group 26. This entry containd a 1927 description of the Nauset Beach lights.

Index to Lighthouse Board Letterbooks (Nauset Beach lights) Records of the United States Coast Guard. Record Group 26. All that remains from the Commerce Department fire of 1921 are the index to these letterbooks. The one or two index sentences describing each letter's contents provide some information, but do not give a full account of each letter's subject matter.

Letter from the Secretary of the Treasury transmitting the Report of the General Superintendent of the Lighthouse Establishment, December 19, 1850. Records of the United States Coast Guard Record Group 26. The report contains a list of supplies delivered to Nauset Beach in 1850.

Lighthouse Superintendent, Boston 1826-44 (correspondence). Records of the United States Coast Guard. Record Group 26. This group of documents has a small amount of material pertaining to the early Nauset Beach lights.

United States Coast Guard Academy, New London, Connecticu

Group Plan Map of Nauset Beach structures. No date.

First District Coast Guard, Boston, Massachusetts

Description of Lighthouse Towers, Buildings, and Premises at Nauset Beach, Massachusetts, Light-Station. January 11, 1910. The document gives a fairly good description of the three lights as to their dimensions including doors and windows.

Cape Cod National Seashore

three
account
what they did

of the Treasury
Printing Office,

of the Treasury
Government Printing

United States, 1873.

United States to the
Ending June 30, 1874.
1874.

Secretary of the Treasury
1876 Washington, D.C.:

the Secretary of the Treasury
1882. Washington, D.C.:

to the Secretary of the Treasury
June 30, 1892. Washington, D.C.:

use Board to the Secretary of the Treasury
June 30, 1895. Washington, D.C.:
1895.

ssioner of Lighthouses to the Secretary of
the Fiscal Year Ending June 30, 1911.
Government Printing Office, 1911.

ort of the Commissioner of Lighthouses to the Secre
ce and for the Fiscal Year Ending June
gton D.C.: Government Printing Office, 1913.

ort of the Commissioner of Lighthouses
ce for the Fiscal Year Ending June

D.C.: Government Printing Office, 1923. Although th run of annual lighthouse reports were reviewed from 1852 to 1925 only the above reports contained a small amount of information on the Nauset Beach lights.

House of Representatives. "Report of I.W.P. Lewis, Civil Engineer, made by Order of Hon. W. Forward, Secretary of the Treasury on the Condition of Light-Houses, Beacons, Buoys, and Navigation, upon the Coasts of Maine, New Hampshire, and Massachusetts, in 1842." Doc. No. 183. February 24, 1843. 27th Cong., 3rd Sess. Lewis's report is very revealing about the condition of the Light-House Establishment under Fifth Auditor Stephen Pleasonton. He singles out the Nauset Beach lights as a horrible example of the conduct of Pleasonton's administration.

House of Representatives. Letter from the Secretary of the Treasury transmitting estimates of appropriations required for the Service of the Fiscal Year Ending June 30, 1854. Ex. Doc. No 2, 32nd Cong., 2nd Sess. This document contains a list of keepers salaries for the period.

United States Statutes At Large. 5. Boston: Charles C. Little and James Brown, 1850.

Secondary Works

Books

Baker, William A. A History of the Boston Marine Society, 1742-1967. Boston: Boston Marine Society, 1968. Baker was commissioned by the Boston Marine Society to compile its history. Although the book is dull and redundant, it does cover the history of that organization quite well and mentions how the Nauset Beach lights came into existence.

Holland, Francis Ross, Jr. America's Lighthouses: Their Illustrated History Since 1716. Brattleboro, Vermont: The Stephen Greene Press, 1972. Holland's work has been accepted as the standard account of the lighthouse service. Factually, it is a good balanced work. His opinions, however, indicate that he lacks a general knowledge of American history.

Johnson, Arnold B. The Modern Light-House Service. Washington, D.C.: Government Printing Office, 1890. Johnson was a Light-House Service employee who was asked to produce a study of his organization. Except for not presenting the controversy over Stephen Pleasonton's administration of the Lighthouse Establishment, Johnson wrote one of the best accounts of the service up to 1889 and can still be considered an excellent book on the early history of

D.C.: Government Printing Office, 1923. Although the run of annual lighthouse reports were reviewed from 1852 to 1925, only the above reports contained a small amount of information on the Nauset Beach lights.

House of Representatives. "Report of I.W.P. Lewis, Civil Engineer, made by Order of Hon. W. Forward, Secretary of the Treasury on the Condition of Light-Houses, Beacons, Buoys, and Navigation, upon the Coasts of Maine, New Hampshire, and Massachusetts, in 1842." Doc. No. 183. February 24, 1843. 27th Cong., 3rd Sess. Lewis's report is very revealing about the condition of the Light-House Establishment under Fifth Auditor Stephen Pleasonton. He singles out the Nauset Beach lights as a horrible example of the conduct of Pleasonton's administration.

House of Representatives. Letter from the Secretary of the Treasury transmitting estimates of appropriations required for the Service of the Fiscal Year Ending June 30, 1854. Ex. Doc. No. 2, 32nd Cong., 2nd Sess. This document contains a list of keepers' salaries for the period.

United States Statutes At Large. 5. Boston: Charles C. Little and James Brown, 1850.

Secondary Works

Books

Baker, William A. A History of the Boston Marine Society, 1742-1967. Boston: Boston Marine Society, 1968. Baker was commissioned by the Boston Marine Society to compile its history. Although the book is dull and redundant, it does cover the history of that organization quite well and mentions how the Nauset Beach lights came into existence.

Holland, Francis Ross, Jr. America's Lighthouses: Their Illustrated History Since 1716. Brattleboro, Vermont: The Stephen Greene Press, 1972. Holland's work has been accepted as the standard account of the lighthouse service. Factually, it is a good balanced work. His opinions, however, indicate that he lacks a general knowledge of American history.

Johnson, Arnold B. The Modern Light-House Service. Washington, D.C.: Government Printing Office, 1890. Johnson was a Light-House Service employee who was asked to produce a study of his organization. Except for not presenting the controversy over Stephen Pleasonton's administration of the Lighthouse Establishment, Johnson wrote one of the best accounts of the service up to 1889 and can still be considered an excellent book on the early history of lighthouses.

Weiss, George. The Lighthouse Service: Its History, Activities and Organization. Baltimore: The Johns Hopkins Press, 1926. Weiss's work is very similar to that of Arnold Johnson. He, of course, continued the story beyond the 1889 date of Johnson's book.

Articles

"Light-House Construction and Illumination." Putnam's Magazine. 8 (August 1856) 198-213. The article gives a general history and description of lighthouse construction and lighting through the ages.

Munroe, Kirk. "From Light to Light: A Cruise of the Armeria Supply Ship." Scribner's Magazine. 20 (October 1896) 460-75. An interesting account of the methods used to supply lighthouses.

Nordhoff, Charles. "The Light-Houses of the United States." Harper's New Monthly Magazine. 48 (March 1874) 465-77. The article discusses the development of the American Lighthouse Service.

Architecture Component Bibliography

Edgerton, Charles E. "The Wire-Nail Association of 1895-96." Political Science Quarterly 12, no. 2:246-47.

This article focuses on the economic, political and legal causes of the rpaid transition from widespread use of cut nails to widespread use of wire nails.

Feilden, Bernard M. Conservation of Historic Buildings. London: Butterworth & Co. Ltd., 1982. The definitive guide to the practice of historic preservation and architectural conservation.

Fleming, John, Hugh Honour, and Nikolaus Pevsner, eds. Penguin Dictionary of Architecture. 3rd ed. New York: Penguin Books, 1980.

General reference for architectural terminology.

Fontana, Bernard L., J. Cameron Greenleaf, et al. "Johnny Ward's Ranch: A Study in Historic Archeology." The Kiva 28 (October-December 1962): 48.

Discussion of the validity of wire and cut nails used for dating archeological contexts.

Gettens, Rutherford J., and George L. Stout. Painting Materials. New York: Dover Publications, 1966.

Standard reference for historical paint research.

Harris, Cyril M., ed. Dictionary of Architecture and Construction. New York: McGraw-Hill Book Co., 1975.

General reference for architectural terminology.

Nelson, Lee H., "Nail Chronology as an Aid to Dating Old Buildings," (American Association for State and Local History Technical Light 48, History News, Vol. 24, No. 11, November, 1968. Nashville, Tennessee).

Discusses evolution in manufacture of nails commonly used in building construction and how to "read" the evidence of manufacturing technology on the nails.

Peterson, Charles E., ed. Building Early America: Contributions toward the History of a Great Industry, Radnor, PA: Chilton Book Co., 1976.

A compilation of papers on early American construction technology presented at a symposium in Philadelphia.

Russell and Erwin Manufacturing Company, Illustrated Catalogue of American Hardware, 1865. Reprint ed. Baltimore: Association for Preservation Technology, 1980.

Comprehensive catalogue of all types of hardware. Used here primarily for terminology and identification.

Timmons, Sharon, ed. Preservation and Conservation: Principles and Practices. Washington: National Trust for Historic Preservation, 1976. Proceedings of a conference by that title.

As the Nation's principal conservation agency, the Department of the Interior has basic responsibilities to protect and conserve our land and water, energy and minerals, fish and wildlife, parks and recreation areas, and to ensure the wise use of all these resources. The Department also has major responsibility for American Indian reservation communities and for people who live in island territories under U.S. administration. NPS D-56, February 1986

CPSIA information can be obtained
at www.ICGtesting.com
Printed in the USA
BVHW08s1023021018
529051BV00020B/1074/P